A Clinical Guide for Cancer Prostate Patient

Narmada Gupta
Editor

A Clinical Guide for Cancer Prostate Patient

Springer

Editor
Narmada Gupta
Urology
Medanta Kidney and Urology Institute
Medanta—The Medicity
Gurugram, India

ISBN 978-981-97-5356-7 ISBN 978-981-97-5357-4 (eBook)
https://doi.org/10.1007/978-981-97-5357-4

This Springer imprint is published by the registered company Springer Nature Singapore Pte Ltd.
The registered company address is: 152 Beach Road, #21-01/04 Gateway East, Singapore 189721, Singapore

Dedicated to all my patients with carcinoma prostate

Foreword

It is with immense pleasure and sense of deep respect I am humbled to write this foreword for the monograph entitled *A Clinical Guide for Cancer Prostate Patient* authored by the esteemed Professor N. Gupta and his accomplished team of authors. I find this book an ideal mix of the state-of-the-art knowledge on the current trends on the subject and rich personal experience of Dr. Gupta for decades in dealing with a wide spectrum of cancer prostate patient. This is also a perfect example of extraordinary collaboration of medical expertise, sharing valuable insights into the intricacies in the management of prostate cancer and presenting as a cohesive wisdom for the benefit of the wide-ranging readership of this monograph.

Dr. Gupta's impeccable record as a distinguished urologist, with longstanding tenure as a Chairman of Urology at the All India Institute of Medical Sciences, New Delhi, and subsequently at Medanta, the Medicity, is the testament to his unparalleled commitment to patient care, high quality of training of the urologist and significant research in the field. This book is yet another example of his relentless pursuit to excellence.

Cancer prostate is a formidable adversary, affecting countless lives around the world. As we navigate complexities of this disease in the context of emerging concepts and treatments emphasis on shared decision-making, this book will serve as beacon of knowledge and understanding of the patient and his family. This collective effort presents comprehensive nature of the contents on the subject ranging from its pathophysiology, variability in the clinical behavior, early detection, prevention, advanced treatment modalities, and supportive care.

At one end, the book offers young professionals to deepen their understanding of this disease in the pursuit of providing the best care to their patients and at the other end a simplified companion to the patients to get answers to their questions.

One of the many strengths of the book lies in their multi-authored approach collaborating with experts from diverse backgrounds and institutions to create this well-structured, researched, and well-rounded evidence-based compendium. The amalgamation of collective experience, perspectives, and clinical acumen within these pages add immense value to the field of urology and prostate cancer research.

In the rapidly evolving landscape of management strategies of cancer prostate, this book meets the challenge of keeping its readership up to date with the latest advancements in the field and covers the contemporary aspects of prostate cancer management as well as offers insights into emerging therapies and newer diagnostic modalities.

The chapters on risk stratification and management of prostate cancer across various risk groups are invaluable. Tailoring treatment options to suit individual patient plan need is crucial and this book offers authentic strategies to achieve that goal. The emphasis on active surveillance as a viable option in well-selected cases reflects growing trends towards personalized care, maximizing outcomes while minimizing treatment inflicted side effects.

The comprehensive coverage of advanced disease management, including hormone-sensitive and castrate-resistant metastatic prostate cancer, is commendable. These chapters offer insights into cutting-edge therapies and emerging treatments trends. The chapter on genetic testing for cancer prostate is a new emerging field that helps for screening of the patient and family members and personalized management in advanced disease. In Chap. 14, various diagnostic and treatment modalities are described in detail for better understanding for the patients to know them before undergoing these procedures.

In conclusion, *A Clinical Guide for Cancer Prostate Patient* is a treasure trove of knowledge, borne out of collective wisdom and experience of leading experts. They are a testament of relentless pursuit of excellence in the field of urology and oncology. Whether you are a seasoned practitioner, a budding urologist, or an anxious patient, this work will undoubtedly enrich your understanding and refine your approach to dealing with prostate cancer in the best possible manner.

I extend my heartfelt appreciation to Professor N. Gupta and all the esteemed co-authors for their invaluable contributions in this seminal work. May this guide continue to inspire progress in prostate cancer care, guiding us towards a future where we conquer this disease and improve the lives of the patients worldwide.

Robotic Surgery Education and Research,
Vattikuti Urology Institute, Henry Ford Hospital,
Vattikuti Foundation, Detroit, MI, USA

Department of Urology, SGPGI, Lucknow, India

Mahendra Bhandari

Preface

Fifty years ago, when I joined the Department of Surgery, Jabalpur Medical College, diagnosis of cancer prostate was by digital rectal examination for hard prostate, sometimes finger-guided transrectal biopsy and treatment was bilateral orchidectomy and if patient in retention of urine, then suprapubic cystostomy. During these years, I have treated hundreds of patients both at AIIMS, a government hospital and Medanta, the Medicity, a corporate hospital. My patients ranged from 36 years to 96 years at the time of diagnosis and from all walks of life including politicians, bureaucrats, diplomats, judges, advocates, doctors, artist, businessman, rich and poor. After discussion, they like to choose an option according to their convenience. In long-term follow-up, the outcomes are same and the patients are satisfied with their quality of life. Shared decision-making is important for the management of cancer prostate.

Due to the increase in longevity of life and awareness both in public and doctors, patients with cancer prostate are increasing in number. There is also an increase in the understanding of aetiopathogenesis, diagnostic tests, imaging and advances in technology for management. As a result, the management of cancer prostate is revolutionised. There is an increase in research and voluminous literature is published every month on cancer prostate. This has also resulted in confusion to decide on diagnosis and management in a particular patient.

The first treatment of the cancer prostate is very important. It should not be under or over treatment. It will decide survival benefits and quality of life after treatment. Patient will also need follow-up for evaluation of recurrence of the disease and further treatment lifelong.

The aim of this monograph is to share the knowledge and guide for proper diagnosis and management of a patient with cancer prostate. The monograph will provide guidelines for evaluation and investigation to reach a diagnosis. After diagnosis, a proper risk stratification of the disease is important to evaluate the risk of patient survival and quality of life after the treatment. According to the risk stratification, treatment options will be analysed. Patient and disease factors decide on appropriate treatment for an individual patient. The outcomes of the treatment are to be analysed for further follow-up treatment.

This monograph is divided into two sections. The first section is about diagnosis, evaluation, pathology, imaging, risk stratification, and management strategies according to the risk of the disease, prevention, and palliative care. In the second section, chapters are included on prostatic biopsy and each modality of treatment for its indications, contraindications, complications and outcomes so that they can be utilised properly. This will help the patient to understand the limitations and outcomes of each modality.

The paragraphs in each chapter are given in bullets to highlight the important factors. This is not a reference/textbook. This book is focussed on practical aspects of management of carcinoma prostate. Summarising this large voluminous literature into paragraphs of few lines was a herculean task for every author and editor.

I hope this monograph will be useful for the patients for understanding of their disease, their expectations and outcomes. Similarly, it will be useful for the post-graduate students and practicing surgeons and urologist to inform the patient about the disease and help them in shared decision-making.

Gurugram, India Narmada Gupta

Acknowledgement

I am grateful to my late parents Shri. Mohanlal and Sato Bai who made me a doctor to serve the humanity. My thanks to thousands of my patients with cancer prostate who helped me gain experience in diagnosis and treatment of this complex disease. During interactions and follow-up, they gave me their side of story about their disease and feelings about the treatment outcomes and quality of life after the treatment. This information inspired me prepare this book *A Clinical Guide for Cancer Prostate Patient* which can help them to understand their disease and participate in decision-making for their personal treatment with the treating doctor. This shared decision is very important for the management of cancer prostate.

I am grateful to all my authors who accepted my request for their contribution to this book. They worked hard and summarised the latest literature available in simple language for understanding of the patient.

My sincere thanks to Dr. Mahender Bhandari, formerly Director, Professor and Head, Department of Urology and Renal Transplantation, Sanjay Gandhi Institute of Medical Sciences; currently Director of Robotic Surgery Education and Research, Vattikuti Urology Institute, Henry Ford Hospital, and CEO of Vattikuti Foundation, Detroit, MI, USA, for critically reviewing and writing the foreword for this book.

I will always be indebted to the All India Institute of Medical Sciences, New Delhi (AIIMS) and Medanta, The Medicity, where I worked and treated these patients during the last 46 years.

I am grateful to my publisher ELSEVIER for publishing the book. I personally thank Dr. Naren Agarwal, Raman Shukla and Jayanthi for their guidance and help in preparing this manuscript.

Last but not least, I am thankful to my wife Dr. Usha, daughter and son-in-law Priyanka and Manish and my grandson Vedant who helped me in my career to fulfil my dreams.

Contents

About the Editor

Narmada Gupta, MBBS, MS, MCh Urology, FAMS, DSc who was born on 12th October 1947, after his medical school and post-graduation in Surgery and Urology, joined as a faculty in the Department of Urology at All India Institute of Medical Sciences, New Delhi, in 1979 and rose to the position of Professor and Chairman from 1998 to 2010. He joined as Chairman, Academic and Research in Urology in 2010–2023 at Medanta, Gurgaon, and is currently working as Professor Emeritus.

During a distinguished career of 46 years, he contributed immensely to the growth of urology in India and Asia. He has published more than 330 scientific papers in various national and international journals. He has written six books on urology; contributed nine chapters and also produced 20 teaching video films. He is a teacher par excellence and has trained over 140 MCh/DNB, post-graduates in Urology.

Dr. Gupta is a pioneer and started several new urologic procedures in India in the field of endo-urology, laparoscopic surgery, laser and robotic surgery. He has performed more than 40,000 urological operations for BPH, carcinoma bladder, stricture urethra, endo-urology, microscopic surgery for male infertility, laser and prosthetic surgery. Under his leadership, Robotic surgery in Urology started in India in 2006. He has performed and popularised Robotic surgery for carcinoma prostate, carcinoma bladder, carcinoma kidney and reconstructive surgery for a variety of urological diseases.

Dr. Gupta served SIU as a member since 1982, as National Delegate, Indian National Section (1994–2000), Board member of SIU (2004–2006), President of the Urological Society of India (2007–2008) and President of the Urological Association of Asia (2008–2010). He was the Organising Chairman of the 9th Asian Congress of Urology in 2008.

He had many awards and honours, notably Dr. B.C. Roy National Award by Medical Council of India for a distinguished teacher (2005), Himadri Sarkar Memorial Oration by USI (2009), Urology Gold Medal by USI (2009), Life Time Achievement Award by North Zone Chapter, USI (2010), Founder Chairman of Robotic Urology Forum, India (2017–2019). He was conferred "Padma Shree" by the Government of India (2007) in recognition of the distinguished services in Urology to the nation.

Chapter 1
Prostate Cancer: Disease Burden

Sahil K. Pethe and Narmada Gupta

Number of cancer prostate cases are increasing due to awareness and diagnostic facilities

- Prostate cancer (PC) is the second most frequently diagnosed cancer in men [1]. It is derived from prostatic epithelial cells and is a highly heterogeneous disease ranging from asymptomatic cases needing active surveillance to highly progressive cases requiring immediate action.
- There is a significant rise in the burden of PC owing to population aging and increased PSA screening [2]. An estimated 1.4 million new cases of PC were diagnosed worldwide in the year 2020. A total of 375,304 men died of PC across the world in 2020 making PC the fifth leading cause of cancer related deaths [3].
- The Human Development Index (HDI) is a measure of the average achievement of human development of each country based on parameters like life expectancy at birth, education index, and gross national per capita income [4]. Positive correlation was found between incidence rate and countries with very high HDI, whereas, Age Standardized Mortality Rates (ASMRs) were inversely proportional to HDI [1].
- There is a huge difference in incidence rates across the world ranging from a low of 6.3 to as high as 83.4 per 100,000 men. Northern and Western Europe, the Caribbean, and Australia/New Zealand have detected higher rates, whereas the lowest incidence is found in Asia and North Africa.
- Age-standardized mortality rates (ASMRs) have decreased recently in 70 out of 89 countries reflecting large scale PSA screening, active surveillance, and improved management of PC [5]. Effects of PSA screening from important trials show an increase in the number of newly diagnosed cases, by 12% in the PLCO

S. K. Pethe
Medanta Kidney and Urology Institute, Medanta—The Medicity, Gurugram, Gurugram, Delhi NCR, India

N. Gupta (✉)
Urology, Medanta Kidney and Urology Institute, Medanta—The Medicity, Gurugram, India

N. Gupta (ed.), *A Clinical Guide for Cancer Prostate Patient*, https://doi.org/10.1007/978-981-97-5357-4_1

(Prostate, Lung, Colorectal, and Ovarian) screening trial [6], 91% in the ERSPC (European Randomized Study of Screening for Prostate Cancer [7], and 51% in the Goteborg Randomized Population based Prostate Cancer Screening Trial [8].

- PC incidence rate in India was 4.8 per 100,000 men in 2016 with an increase in ASIR by 29.8% from 1990 to 2016 [9]. The exact incidence rate of PC in India is limited due to very few community-based studies on PC [10, 11]. There is an increasing trend in the diagnosis of new cases of PC in men ≥65 years of age according to the various state registries of cancer in India [12]. Almost 85% of PC patients detected annually are in the late stages (stage III and IV) as compared to the US (15%) [13].
- Developing countries such as India have more cases in advanced stages as compared to developed nations due to resource limitations, risk of false positive results, and treatment complications.
- 5-year survival rates for PC are relatively low in Asian countries (61%) as compared to Europe and North America (83% and 98%, respectively) [14–16].

1.1 Etiopathogenesis

- Prostate cancer is the leading cause of extracutaneous cancer in males. The aetiology of prostate cancer cannot be pinpointed as there are multiple risk factors associated with it. Although some established risk factors are known which are non-modifiable like old age, positive family history, and African origin (Table 1.1).

- **Age**: The incidence of prostate cancer increases drastically with age and maximum are diagnosed after the age band of 60 years [5]. The prevalence is observed to triple in the sixth to ninth decade [17]. There is documented histological evidence indicating an increased probability of developing PC in higher age groups. The probability percentages are 0.005% in <39 years, 2.2% in 40–59 years, 13.7% in 60–79 years, and 50% in 70–80 years old men [18]. The GLOBACON study found an increased incidence of PC in all age bands due to advanced screening methods and thus early detection of prostate cancers. However, these cases are mostly asymptomatic [5].

Table 1.1 Risk factors for prostate cancer

Modifiable	Non-modifiable
• Hormones	• Age
• Diet	• Family history and genetic factors
• Smoking, alcohol	• Race/ethnicity
• Physical activity/lifestyle	
• Environmental agents	
• Chronic inflammation	
• Others	

- **Family history and genetic factors**: PC has a significant familial association. The risk of occurrence is directly proportional to the number of relatives with PC and the early age of diagnosis. There is a two to threefold increase if it is a first-degree relative and more so if it is a sibling diagnosed with PC [17]. Some studies have observed an increased risk of PC in families with familial breast cancer traits.
- A US-based large prospective study had identified a 21% increased risk of PC in men with familial breast cancer [19]. The familial links of PC suggest a strong association with PC causing genes.
- Genome Wide Association Studies (GWAS) are being conducted for identifying the genetic predisposing factors of PC. A total of 167 prostate cancer susceptibility loci have been known till now after the largest PC GWAS and reporting of rarer single nucleotide variants in some studies [20].
- The earlier linkage studies in prostate cancer families identified 8q24 as an important PC risk region and missense G84E variant in HOXB13. The G84E variant was reported mostly in the European population with the highest rate of 22% in Finland, followed by Sweden (8.2%) according to the International Consortium of Prostate Cancer Genetics (ICPCG) study [21].
- The increased risk of PC in families with familial breast cancer points towards the possibility of inherited BRCA mutations. Even though family-based linkage studies identified certain rare gene variants such as G84E, HOXB13, or BRCA2 but still there are unexplained familial risks.
- The introduction of GWAS has greatly increased the cancer variants' identification power. The first GWAS was reported in 2006 and it identified Single Nucleotide Polymorphisms (SNPs) in chromosome 8q24 [22].
- MYC gene activation was demonstrated in up to 60% of prostate cancer cases even in pre-invasive stages. PTEN loss along with MYC increase correlates with metastasis, poor prognosis, and higher Gleason score [23, 24].
- The translocations in androgen-regulated promoters and the (E26 transformation-specific) ETS family of transcription factors are the most common genetic variations in prostate cancer [25].
- Studies have shown that genomic amplification or mutation in the androgen signalling pathway such as AR-V7, and FOXA1 promotes tumour growth.
- Alterations in TGF-B (tumour suppressor) pathway lead to cancer progression.
- GWAS is advantageous in risk stratification, to target the PC screening of men with increased genetic predisposition [26]. There are no highly penetrant or dominant mutations in prostate cancers like other cancers such as breast cancer. Hence, genetic analysis has a wide scope in future therapeutic inventions.
- **Race/ethnicity**: PC incidence shows marked variations in different racial groups with the highest incidence in black men of African descent. Oriental men from Japan have the lowest risk and Caucasians have a moderate risk of developing PC. The variation in genetic composition across ethnic groups is responsible for such varied incidence rates.
- Diminished activity of protective tumour suppressors due to lower TMPRSS2-ERG fusion, PTEN deletion, and up-regulation of MNX1 in men of African descent promotes cancer development. DARC, the chemokine receptor which removes chemokines from the prostate tumour microenvironment is found to be

decreased in many African men, thus increasing incidence and mortality rates in them [17, 27].

- **Hormones**: Androgens are responsible for the growth and development of the prostate gland. Conversion of testosterone to dihydrotestosterone (DHT) is essential for Androgen receptor (AR) signalling which in turn leads to proliferation of cells and functional maintenance. Hence, it was earlier implicated that higher testosterone levels promote prostate cancer and anti-androgen therapy was one of the foremost treatment options in PC management. However, recent studies show a decline in the association of prostate cancer with high testosterone [28].
- Studies have also observed that higher levels of testosterone do not increase the risk of cancer progression [29–32]. Hypo gonadic individuals who are on testosterone replacement therapies (TRT) do not show an increased risk of PC [33].
- Oestrogen and their mimics have been associated with PC pathogenesis and the progression of cancer according to the findings of several studies. Oestrogens cause hormone induced cell proliferation, cease tumour-suppressor genes, and, activate genotoxic carcinogens. The effective drug therapy of PC by Selective Oestrogen Receptor Modulators (SERMs), phytoestrogens, and other anti oestrogens further confirms the role of oestrogens in the development of PC [34].
- **Diet**: A variety of dietary factors have been associated with prostate cancer. These include animal fat, calcium, dairy products, selenium, processed foods, etc. consumption. The strong correlation between fats, especially polyunsaturated fat, and PC incidence is established in various studies. Mechanisms of action include either fat induced hormonal alterations, production of protein or DNA reactive intermediates, or increased oxidative stress [35].
- Intake of calcium is postulated to suppress calcitriol affecting the cell cycle of prostatic epithelial cells and thus contributing to the development of PC. Dairy products show inconclusive evidence of association with PC [36]. However, some studies have found that the intake of dairy products increases levels of insulin like growth factors which are linked to PC [37].
- The SELECT trial concluded that selenium and vitamin E do not have beneficial effects and rather increase the risk of high-grade cancer [38].
- Consumption of soy foods and lycopene (a red pigmented carotenoid found in tomatoes and watermelons) have shown a significant reduction in the risk of prostate cancer [39, 40].
- Certain studies have shown that consumption of fish oil (vitamin A source) reduces PC risk [17] whereas some show mixed results for fish intake and PC progression [41, 42].
- Increased obesity leads to increased insulin like growth factor 1 (IGF-1), secretion of chemokines CCL7, and fall in adiponectin (an adipokine) ultimately contributing to a greater risk of PC using the anti-inflammatory, anti-proliferative, signalling pathways [43–45].
- Metabolic syndrome is another important risk factor for PC. The REDUCE trial highlighted that the presence of 3 or more components (hypertension, hyper

glycemia, hyper lipidaemia, and excess body fat with increased waist circumference) of metabolic syndrome has a greater risk of high-grade PC [46].

- **Smoking and alcohol**: Smoking and alcohol have been implicated in many cancers. Huncharek et al. [47] and other meta-analyses have shown that the risk of PC increases with an increasing amount of smoking and increasing volume of alcohol intake compared to non-smokers and non-drinkers. Additional findings were that ex-smokers had increased PC risk and heavy smokers had a 24–30% increase in mortality rates [48].
- **Physical activity/lifestyle**: A sedentary lifestyle has a higher risk of progression and PC mortality. Studies have shown a significant reduction in risk of PC mortality by up to 61% [49] in men working out for 3 h a week compared to those who did less than an hour per week.
- **Environmental agents**: Environmental agents such as endocrine-disrupting chemicals (EDCs) affect hormone activities affecting reproduction, growth, and/or carcinogenesis. EDCs can be exposed to humans through food ingestion, water consumption, or inhalation. Studies have identified EDCs in pesticide residues on food, chemicals used in plastic production, phytoestrogens in plant products, and in water downstream of pulp mills.
- The active metabolite of the pesticide vinclozolin is an androgen antagonist which decreases androgen regulated gene expression.
- Cadmium occurring naturally in tiny amounts in the environment is a known carcinogenic agent and is linked to PC in some studies [35].
- **Chronic inflammation**: Cancer related inflammation causes leukocyte migration, cytokine production, chemokine production, angiogenesis, epithelial-mesenchymal transition (EMT), migration, and metastasis. A similar chronic inflammation pathway is observed in prostate cancer also [50].
- PCPT and SELECT studies [51, 52] have shown an association between chronic inflammation and the development of PC. Chemokines produced in PC and other inflammatory milieu interact with the tumour microenvironment leading to the progression of cancer. Important chemokines in PC responsible for its increased incidence and mortality are CXCR-upregulated, and decreased DARC receptor activity.
- Exogenous stimuli such as infections, radiation, hormones, and chemicals play a role in chronic inflammation [27].
- **Others**: Many other factors such as marital status, vasectomy, anthropometry, industrial exposure, etc. with inconsistent risk association are mentioned in some studies. The number of children and vasectomy did not contribute to any significant risk.

1.2 Pathogenesis

- The pathogenesis of any disease plays a major role in deciding the management of the disease. Prostate cancer is a significant medical problem for affected men despite medical advancement over the decades [26].
- The unpredictable nature and inconsistent clinical behaviour of PC are due to its certain characteristics:
- PC is a slow growing cancer with a significantly longer doubling time of 3–4 years. There is a long latency period due to this which contributes to a 10 times higher prevalence rate than incidence.
- PC is typically found in aged men and rarely in men below 40 years of age. This age relatedness is presumed to be due to the accumulation of genetic damage over the years. These genetic alterations may be the result of oxidative stress or other factors.
- Prostate cancers and PIN (Prostatic Intraepithelial Neoplasia) are usually multifocal.
- PC and PIN are heterogenous pathologically suggesting the involvement of multiple pathways and multiple mechanisms for its development.
- The prostate gland is divided anatomically into 3 zones-peripheral (60–75% prostatic cancer arises here), central zone, and transition zone [26]. In 70% of cases cancer arises in peripheral zone, 25% in central and 5% in transition zone.
- Benign hyperplasia of prostate (BPH) and cancer prostate both can occur in prostate gland as both are age related, but BPH do not play role in aetiology for cancer prostate.
- The prostate gland is a tubule-alveolar gland with stromal (fibromuscular) and epithelial (glandular) tissue [17]. The luminal, basal, and neuroendocrine cell types form the epithelial prostate. PC primarily involves epithelium (carcinoma) and rarely involves stromal tissue (sarcoma) [53].
- The luminal cells are on the innermost side of the prostatic ducts; secrete prostatic fluid with glycoprotein PSA and express androgen receptors, cyto keratins (CK8, CK18). The basal cells are responsible for the formation of acinar basement membrane and express high levels of CK5 and p63 and very low AR. Neuroendocrine cells are located on the basal cell layer and express neuroendocrine markers such as synaptophysin and chromogranin A. They do not express ARs and are thus androgen independent [26, 54].
- The transformation of benign epithelial cells into malignant types in the prostate is termed PIN. The PIN is the initial phase of PC which progresses to localized cancer and then advanced adenocarcinoma with local invasion and metastasis in the end [26]. Mechanisms which lead to prostate cancer are uncertain and several theories are proposed to understand it (Table 1.2).

Table 1.2 Theories of prostate cancer pathogenesis

• Role of androgen receptor
• Androgen receptor bypass pathway
• Tumour suppressor genes
• Oxidative stress
• Tumour microenvironment
• Apoptosis regulation
• Epithelial mesenchymal transition

1.3 Androgen Receptor and Prostate Cancer

- ARs play a vital role in prostate gland growth and maintenance. AR activation by DHT helps AR to bind with androgen response elements (AREs) in the DNA and finally takes part in initiating gene transcription which is essential for cell growth, proliferation, and apoptosis [55]. Hence, blocking the AR activation through LBP, 5α-reductase antagonists, or HPG overstimulation via LH/GnRH analogues causes cell apoptosis and thus a treatment option to prevent the proliferation of PC.
- Several studies have linked AR mutations with PC, especially CRPC (Castration resistant PC) and androgen-independent PC [56]. 159 of 1029 reported AR mutations in the database are related to increased risk of PC in men [57].

1.4 Androgen Receptor Bypass Pathway

- Reduced activation of AR causes hypersensitization of other pathways such as upregulation in IGF-1, EGF, and other growth factors. The tyrosine receptor kinases (RTKs) are activated due to these growth factors and lead to stimulation of phosphatidyl inositol 3 kinase (PI3K) and subsequent PI3K-AKT-mTOR pathway [58, 59]. AKT (protein kinase B) deregulates the inhibitor of GTP binding protein RHEB and activates the mechanistic target of rapamycin complex 1 (mTORC1) which stops autophagy and promotes prostate cancer cellular proliferation. This PI3K-AKT-mTOR pathway can have future therapeutic targets for patients with CRPC [58].

1.5 Androgen Receptor Independent Pathway

- Various peptide growth factors help in AR regulated prostate epithelial cell multiplication via a process called "cross talk" [60]. Epidermal growth factor (EGF) and its associated tyrosine receptor kinas EGF-1 increase the invasiveness of cancer cells by migrating through extracellular matrix barriers and basal membranes [61]. ErbB2 (HER2) of the epidermal growth factor receptor family is found to activate ARs independent of androgens and increase the expression of PSA [62].

1.6 Tumour Suppressor Genes

- The initial sign of PC is inflammation of the prostate gland due to uncontrolled cell division which is caused by genetic alterations. The tumour suppressor genes are not found to have individual responsibility for cancer initiation or progress. However, several gene activities such as MYC overexpression, PTEN inactivation, NKX3.1 down regulation, and TMPRSS2-ERG gene fusion contribute to PC development. TMPRSS2-ERG gene fusion activates the ERG oncogenic pathway ultimately resulting in resistance to ADT and induction of the EZH2 mediated PC cell proliferation [63, 64]. ERG fusion gene is detected in around 50% of PC cases [63]. PTEN gene mutations are detected in 20% cases of CRPC tumours after their genetic analysis [65]. PTEN activation halts cellular proliferation by negatively regulating the PI3K-AKT-mTOR pathway. The loss of PTEN results in increased cellular proliferation, AR expression, and reduced apoptosis [62].

1.7 Oxidative Stress

- Oxidative stress is the excessive production of reactive oxygen species (ROS) as well as other oxidants like reactive lipid species and reactive nitrogen species due to an imbalance between oxidant and reductant molecules. These excess ROS cause oxidative DNA damage which is considered an important stimulus in the pathogenesis of several diseases including PC [66].
- The free radicals can also cause neoplasia and metastases by affecting the physiological cellular functions such as cellular proliferation, remodelling, apoptosis, etc. [66].
- Oxidative stress can also lead to chromosomal abnormalities and oncogene activation which promote cancer. ROS include superoxide radicals, hydrogen peroxide, hydroxyl radicals, and singlet oxygen. These are mainly produced by mitochondria in both physiological and pathological conditions such as delayed p53 recruitment, inflammation, etc. The redox imbalance is more likely due to reduced activity of the antioxidant defence systems. The enzymatic components, such as superoxide dismutase (SOD), catalase (CAT), and glutathione peroxidase (GPx) form the defence system against ROS [67]. Androgen receptor activation is shown to promote ROS production [68].
- Some studies have demonstrated reduced glutathione peroxidase expression in prostate tissue in men PC. It is also proposed to create a prooxidant state leading

to genetic alterations and thus leading to malignant changes [69]. MDA (malondialdehyde) is a highly reactive aldehyde whose DNA damage is linked with mutagenic, genotoxic, and cytotoxic effects. MDA is used as an OS measuring parameter and is found to be significantly increased in PC patients in several studies [70–72].

1.8 Tumour Microenvironment

- The prostatic tumour microenvironment constitutes numerous supporting cells such as immune cells, fibroblasts, adipocytes, microvasculature, and extracellular matrix. This complex network forms an ecosystem for the transition of normal cells to cancer cells, the development and growth of cancer cells, and the migration of cancer cells to form metastases [45, 73, 74].
- Aging effects of the microenvironment include increased inflammation, structural changes, collagen network changes, and vascular remodelling which can thus affect PC growth and progression [75–77].
- Some studies suggest that tumour cells infiltrate immune cells to form tumour-associated macrophages (TAMs). This argument was based on the finding that CD8-T cells expressed PSA in aggressive forms of PC. Lymphatic and bone metastases can be enhanced due to this ability [78].
- Some studies have observed that the prostate microenvironment tries to maintain normal organ function and hence can work as a controlling force to inhibit excessive cancerous growth [79, 80].

1.9 Apoptosis Regulation

- Apoptosis is programmed cell death and its dysregulation leads to exponential cell growth which is an important aspect of prostate cancer [81]. The initiation pathways of apoptosis are the intrinsic (mitochondrial) and extrinsic (death receptor) pathways which lead to a common end pathway or execution phase [82].
- The extrinsic pathway begins when death receptors such as type 1 Tumour Necrosis Factor receptor (TNFR1) or Fas protein (CD95) bind with their respective ligand i.e. TNF and Fas ligand (FasL). These death receptors have TNF receptor–associated death domain (TRADD) and Fas associated death domain (FADD), along with caspase 8 [83]. Caspase 8 is the apoptosis initiator caspase [84].
- The intrinsic pathway involves the activation of mitochondria and begins after some intracellular damage such as DNA damage [81]. This pathway is closely regulated by the Bcl-2 family of proteins. The bcl-2 family consists of pro-apoptotic proteins (e.g. Bax, Bad, Bcl-Xs, etc.) and anti-apoptotic proteins (e.g. Bcl-2, Bcl-Xl, Bcl-W, etc.) [85]. The pro-apoptotic proteins induce the

mitochondrial release of cytochrome c which forms apoptosome (apoptosis inducing complexes) along with procaspase 9 and apoptotic protease activating factor-1 (Apaf-1). The apoptosome with an extrinsic pathway activates caspase 3 [86]. Bcl-2 upregulation and increased anti-apoptosis are found in aggressive PC [87].

- The role of apoptosis in the pathogenesis of PC is highlighted by increased expression of Bcl-2 receptors and p53 tumour suppressor gene mutations [88]. Some studies have reported overexpression of Bcl-2 in chemoradiotherapy-resistant phenotypes of PC. This can be used as an important prognostic indicator of prostate cancer [81, 87].

1.10 Epithelial Mesenchymal Transition (EMT)

- EMT plays a major role in the metastasis of PC, especially bone metastasis. The first site of metastases of PC is the draining Lymph nodes, and then to distant organs such as bones, liver, and lungs [89]. Prostate cells undergo Epithelial mesenchymal transition (EMT) and thus cross physical barriers to enter the circulation and travel through bone marrow stroma to reach the endosteal bone surface [90]. These circulating tumour cells (CTCs) have vast heterogeneity and are being studied for metastasis and identifying biomarkers [91]. Stromal cell-derived factor 1 (SDF-1 or CXCL 12), its receptor CXCR4 and its anchor Annexin A2 (ANXA2) have been linked to bone metastasis of PC [92]. Annexin A2 is also associated with increased expression in PC cells and it enhances proliferation and inhibits apoptosis during chemotherapy [93]. PC cells also express an adhesion molecule integrin $\alpha v\beta 3$ which binds RGD peptide on ECM proteins to enhance bone endosteum invasion [94]. Once PC cells enter the bone marrow and colonize it, several dynamic changes occur due to the interaction of cancer cells, osteoclasts, and osteoblasts in the bone microenvironment resulting in bone remodelling [26].
- The progression of the disease will have a benign course or rapid metastasis with fatal consequences is still not fully understood and there is a lot of scope for novel innovations for better understanding and management of prostate cancer.

1.11 Burden in India

- With the increase in longevity of life, education, awareness, and availability of PSA testing, the number of patients of cancer prostate are increasing every year. There is also stage migration and early stage cases in younger population are being diagnosed.
- In various cancer registries in India, incidence of cancer prostate was fifth but in latest publications in some registries, it is second most common cancer in men.

However, the data regarding the true incidence of prostate cancer is limited mainly due to the fact that this entity is not a notifiable disease and there are only a handful of population-based cancer registries in India. The number of community-based studies on prostate cancer is also limited in this country.
- Meticulous implementation of registration and reporting mechanisms may improve knowledge regarding prostate cancer in India.

References

1. Wang L, Lu B, He M, Wang Y, Wang Z, Du L. Prostate cancer incidence and mortality: global status and temporal trends in 89 countries from 2000 to 2019. Front Public Health. 2022;10:811044. https://doi.org/10.3389/fpubh.2022.811044.
2. Culp MB, Soerjomataram I, Efstathiou JA, Bray F, Jemal A. Recent global patterns in prostate cancer incidence and mortality rates. Eur Urol. 2020;77:38–52. https://doi.org/10.1016/j.eururo.2019.08.005.
3. Sedláčková H, Dolejšová O, Hora M, Ferda J, Hes O, Topolčan O, Fuchsová R, Kučera R. Prostate cancer diagnostic algorithm as a "road map" from the first stratification of the patient to the final treatment decision. Life. 2021;11(4):324. https://doi.org/10.3390/life11040324.
4. United Nations Development Programme. Human Development Report 2020. The next frontier human development and the Anthropocene. http://hdr.undp.org/sites/default/files/hdr2020.pdf.
5. Sung H, Ferlay J, Siegel RL, Laversanne M, Soerjomataram I, Jemal A, et al. Global cancer statistics 2020: GLOBOCAN estimates of incidence and mortality worldwide for 36 cancers in 185 countries. CA Cancer J Clin. 2021;71:209–49.
6. Andriole GL, Crawford ED, Grubb RL 3rd, Buys SS, Chia D, Church TR, et al. Prostate cancer screening in the randomized prostate, lung, colorectal, and ovarian cancer screening trial: mortality results after 13 years of follow-up. J Natl Cancer Inst. 2012;104:125–32. https://doi.org/10.1093/jnci/djr500.
7. Schröder FH, Hugosson J, Roobol MJ, Tammela TL, Zappa M, Nelen V, et al. Screening and prostate cancer mortality: results of the European Randomised Study of Screening for Prostate Cancer (ERSPC) at 13 years of follow-up. Lancet. 2014;384:2027–35. https://doi.org/10.1016/S0140-6736(14)60525-0.
8. Hugosson J, Godtman RA, Carlsson SV, Aus G, Grenabo Bergdahl A, Lodding P, et al. Eighteen-year follow-up of the Göteborg Randomized Population-based Prostate Cancer Screening Trial: effect of sociodemographic variables on participation, prostate cancer incidence and mortality. Scand J Urol. 2018;52:27–37. https://doi.org/10.1080/21681805.2017.1411392.
9. Dhillon PK, Mathur P, Nandakumar A, Fitzmaurice C, Kumar GA, Mehrotra R, et al. The burden of cancers and their variations across the states of India: the Global Burden of Disease Study 1990–2016. Lancet Oncol. 2018;19:1289–306.
10. Hariharan K, Padmanabha V. Demography and disease characteristics of prostate cancer in India. Indian J Urol. 2016;32:103–8.
11. Jain S, Saxena S, Kumar A. Epidemiology of prostate cancer in India. Meta Gene. 2014;2:596–605.
12. Budukh A, Bakshi G, Prakash G. Change in the ranking and increasing trend of the prostate cancer from the population based cancer registries in India. Indian J Urol. 2018;34:235–6.
13. Hebert JR, Ghumare SS, Gupta PC. Stage at diagnosis and relative differences in breast and prostate cancer incidence in India: comparison with the United States. Asian Pac J Cancer Prev. 2006;7:547–55.

14. Hassanipour S, Delam H, Arab Zozani M, Abdzadeh E, Hosseini SA, Nikbakht HA, et al. Survival rate of prostate cancer in Asian countries: a systematic review and meta analysis. Ann Glob Health. 2020;86:2.
15. Emanuele C. European Commission. Epidemiology of prostate cancer in Europe. 2015. https://ec.europa.eu/jrc/en/publication/epidemiologyprostatecancereurope.
16. American Cancer Society. Survival rates for prostate cancer. 2020. https://www.cancer.org/cancer/prostatecancer/detectiondiagnosisstaging/survivalrates.html.
17. Gleave M, Bandyk M, Chung L. Chapter 12: Prostate cancer. In: Some aspects of oncology; 1996. p. 257–91. https://doi.org/10.1016/s1569-254x(96)80014-8.
18. Newcomer LM, Stanford JL, Blumenstein BA, Brawer MK. Temporal trends in rates of prostate cancer: declining incidence of advanced stage disease, 1974–1994. J Urol. 1997;158(4):1427–30. https://doi.org/10.1016/S0022-5347(01)64231-9.
19. Barber L, Gerke T, Markt SC, Peisch SF, Wilson KM, Ahearn T, et al. Family history of breast or prostate cancer and prostate cancer risk. Clin Cancer Res. 2018;24(23):5910–7. https://doi.org/10.1158/1078-0432.CCR-18-0370.
20. Schumacher FR, Al Olama AA, Berndt SI, Benlloch S, Ahmed M, Saunders EJ, et al. Association analyses of more than 140,000 men identify 63 new prostate cancer susceptibility loci. Nat Genet. 2018;50(7):928–36. https://doi.org/10.1038/s41588-018-0142-8.
21. Xu J, Lange EM, Lu L, Zheng SL, Wang Z, Thibodeau SN, et al. HOXB13 is a susceptibility gene for prostate cancer: results from the International Consortium for Prostate Cancer Genetics (ICPCG). Hum Genet. 2013;132:5–14.
22. Benafif S, Kote-Jarai Z, Eeles RA, Consortium P. A review of prostate cancer genome-wide association studies (GWAS). Cancer Epidemiol Biomark Prev. 2018;27(8):845–57. https://doi.org/10.1158/1055-9965.EPI-16-1046.
23. Weiner HL, Kuchroo VK. Multiple sclerosis. Cold spring harbor perspectives in medicine, vol. 94. New York: Cold Spring Harbor Laboratory Press; 2019. p. 450. https://doi.org/10.1086/706427.
24. Jamaspishvili T, Berman DM, Ross AE, Scher HI, De Marzo AM, Squire JA, Lotan TL. Clinical implications of PTEN loss in prostate cancer. Nat Rev Urol. 2018;15:222–34.
25. Sizemore GM, Pitarresi JR, Balakrishnan S, Ostrowski MC. The ETS family of oncogenic transcription factors in solid tumours. Nat Rev Cancer. 2017;17:337–51.
26. Wang G, Zhao D, Spring DJ, DePinho RA. Genetics and biology of prostate cancer. Genes Dev. 2018;32(17–18):1105–40. https://doi.org/10.1101/gad.315739.118.
27. Rani A, Dasgupta P, Murphy JJ. Prostate cancer: the role of inflammation and chemokines. Am J Pathol. 2019;189(11):2119–37. https://doi.org/10.1016/j.ajpath.2019.07.007.
28. Boyle P, et al. Endogenous and exogenous testosterone and the risk of prostate cancer and increased prostate-specific antigen (PSA) level: a meta-analysis. BJU Int. 2016;118(5):731–41. https://doi.org/10.1111/bju.13417.
29. Lane BR, et al. Low testosterone and risk of biochemical recurrence and poorly differentiated prostate cancer at radical prostatectomy. Urology. 2008;72(6):1240–5. https://doi.org/10.1016/j.urology.2008.06.001.
30. Lee JK, et al. Preoperative serum sex hormone-binding globulin level is an independent predictor of biochemical outcome after radical prostatectomy. Medicine (Baltim). 2015;94(28):e1185. https://doi.org/10.1097/MD.0000000000001185.
31. Roder MA, et al. Serum testosterone level as a predictor of biochemical failure after radical prostatectomy for localized prostate cancer. BJU Int. 2012;109(4):520–4. https://doi.org/10.1111/j.1464-410X.2011.10335.x.
32. Yamamoto S, et al. Preoperative serum testosterone level as an independent predictor of treatment failure following radical prostatectomy. Eur Urol. 2007;52(3):696–701. https://doi.org/10.1016/j.eururo.2007.03.052.
33. Golla V, Kaplan AL. Testosterone therapy on active surveillance and following definitive treatment for prostate cancer. Curr Urol Rep. 2017;18(7):49. https://doi.org/10.1007/s11934-017-0695-6.

34. Ho M, Lam M, Leung K. Estrogens and prostate cancer: etiology, mediators, prevention, and management. Endocrinol Metab Clin N Am. 2011;40(3):591–614. https://doi.org/10.1016/j.ecl.2011.05.002.
35. Bostwick DG, Burke HB, Djakiew D, Euling S, Ho SM, Landolph J, Morrison H, Sonawane B, Shifflett T, Waters DJ, Timms B. Human prostate cancer risk factors. Cancer. 2004;101(10 Suppl):2371–490. https://doi.org/10.1002/cncr.20408.
36. Wiseman M. The second World Cancer Research Fund/American Institute for Cancer Research expert report. Food, nutrition, physical activity, and the prevention of cancer: a global perspective. Proc Nutr Soc. 2008;67(3):253–6. https://doi.org/10.1017/S002966510800712X.
37. Chan JM, Stampfer MJ, Ma J, Gann P, Gaziano JM, Pollak M, et al. Insulin-like growth factor-I (IGF-I) and IGF binding protein-3 as predictors of advanced-stage prostate cancer. J Natl Cancer Inst. 2002;94(14):1099–106. https://doi.org/10.1093/jnci/94.14.1099.
38. Lippman SM, Klein EA, Goodman PJ, Lucia MS, Thompson IM, Ford LG, et al. Effect of selenium and vitamin E on risk of prostate cancer and other cancers: the selenium and vitamin E cancer prevention trial (SELECT). JAMA. 2009;301(1):39–51. https://doi.org/10.1001/jama.2008.864.
39. Rowles JL, Ranard KM, Smith JW, An R, Erdman JW. Increased dietary and circulating lycopene are associated with reduced prostate cancer risk: a systematic review and meta-analysis. Prostate Cancer Prostatic Dis. 2017;20(4):361–77. https://doi.org/10.1038/pcan.2017.25.
40. Applegate CC, Rowles JL, Ranard KM, Jeon S, Erdman JW. Soy consumption and the risk of prostate cancer: an updated systematic review and meta-analysis. Nutrients. 2018;10(1):40. https://doi.org/10.3390/nu10010040.
41. Chan JM, Holick CN, Leitzmann MF, Rimm EB, Willett WC, Stampfer MJ, et al. Diet after diagnosis and the risk of prostate cancer progression, recurrence, and death (United States). Cancer Causes Control. 2006;17(2):199–208. https://doi.org/10.1007/s10552-005-0413-4.
42. Richman EL, Stampfer MJ, Paciorek A, Broering JM, Carroll PR, Chan JM. Intakes of meat, fish, poultry, and eggs and risk of prostate cancer progression. Am J Clin Nutr. 2010;91(3):712–21. https://doi.org/10.3945/ajcn.2009.28474.
43. Smith LA, O'Flanagan CH, Bowers LW, Allott EH, Hursting SD. Translating mechanism-based strategies to break the obesity-cancer link: a narrative review. J Acad Nutr Diet. 2018;118(4):652–67. https://doi.org/10.1016/j.jand.2017.08.112.
44. Liao Q, Long C, Deng Z, Bi X, Hu J. The role of circulating adiponectin in prostate cancer: a meta analysis. Int J Biol Mark. 2015;30(1):e22–31. https://doi.org/10.5301/jbm.5000124.
45. Welén K, Damber E. Androgens, aging, and prostate health. Rev Endocr Metab Disord. 2022;23(6):1221–31. https://doi.org/10.1007/s11154-022-09730-z.
46. Sourbeer KN, Howard LE, Andriole GL, Moreira DM, Castro-Santamaria R, Freedland SJ, et al. Metabolic syndrome-like components and prostate cancer risk: results from the reduction by Dutasteride of Prostate Cancer Events (REDUCE) study. BJU Int. 2015;115(5):736–43. https://doi.org/10.1111/bju.12843.
47. Huncharek M, Haddock KS, Reid R, Kupelnick B. Smoking as a risk factor for prostate cancer: a meta-analysis of 24 prospective cohort studies. Am J Public Health. 2010;100(4):693–701. https://doi.org/10.2105/AJPH.2008.150508.
48. Zhao J, Stockwell T, Roemer A, Chikritzhs T. Is alcohol consumption a risk factor for prostate cancer? A systematic review and meta-analysis. BMC Cancer. 2016;16(1):845. https://doi.org/10.1186/s12885-016-2891-z.
49. Kenfield SA, Stampfer MJ, Giovannucci E, Chan JM. Physical activity and survival after prostate cancer diagnosis in the health professionals follow-up study. J Clin Oncol. 2011;29(6):726–32. https://doi.org/10.1200/JCO.2010.31.5226.
50. Hughes CE, Nibbs RJB. A guide to chemokines and their receptors. FEBS J. 2018;285(16):2944–71. https://doi.org/10.1111/febs.14466.
51. Thompson IM, et al. The influence of finasteride on the development of prostate cancer. N Engl J Med. 2003;349(3):215–24. https://doi.org/10.1056/NEJMoa030660.

52. Klein EA, et al. Vitamin E and the risk of prostate cancer: the selenium and vitamin E cancer prevention trial (SELECT). JAMA. 2011;306(14):1549–56. https://doi.org/10.1001/jama.2011.1437.
53. Humphrey PA. Histopathology of prostate cancer. Cold Spring Harb Perspect Med. 2017;7(10):a030411. https://doi.org/10.1101/cshperspect.a030411.
54. Lee SH, Shen MM. Cell types of origin for prostate cancer. Curr Opin Cell Biol. 2015;37:35–41. https://doi.org/10.1016/j.ceb.2015.10.002.
55. Hodges C. Studies on prostatic cancer I. The effect of castration, of oestrogen and of androgen injection on serum phosphatases in metastatic carcinoma of the prostate. Cancer Res. 1941;1(4):293–7.
56. Taplin M-E, Bubley GJ, Shuster TD, Frantz ME, Spooner AE, Ogata GK, et al. Mutation of the androgen-receptor gene in metastatic androgen-independent prostate cancer. N Engl J Med. 1995;332(21):1393–8. https://doi.org/10.1056/NEJM199505253322101.
57. Gottlieb B, Beitel LK, Nadarajah A, Paliouras M, Trifiro M. The androgen receptor gene mutations database: 2012 update. Hum Mutat. 2012;33(5):887–94. https://doi.org/10.1002/humu.22046.
58. Sarker D, Reid AHM, Yap TA, de Bono JS. Targeting the PI3K/AKT pathway for the treatment of prostate cancer. Clin Cancer Res. 2009;15(15):4799–805. https://doi.org/10.1158/1078-0432.CCR-08-0125.
59. Shorning BY, Dass MS, Smalley MJ, Pearson HB. The PI3K-AKT-mTOR pathway and prostate cancer: at the crossroads of AR, MAPK, and WNT signalling. Int J Mol Sci. 2020;21(12):4507. https://doi.org/10.3390/ijms21124507.
60. Zhu ML, Kyprianou N. Androgen receptor and growth factor signalling cross-talk in prostate cancer cells. Endocr Relat Cancer. 2008;15(4):841–9. https://doi.org/10.1677/ERC-08-0084.
61. Wells A, Kassis J, Solava J, Turner T, Lauffenburger DA. Growth factor-induced cell motility in tumour invasion. Acta Oncol. 2002;41(2):124–30. https://doi.org/10.1080/028418602753669481.
62. Heinlein CA, Chang C. Androgen receptor in prostate cancer. Endocr Rev. 2004;25(2):276–308. https://doi.org/10.1210/er.2002-0032.
63. Wasim S, Lee S-Y, Kim J. Complexities of prostate cancer. Int J Mol Sci. 2022;23:14257. https://doi.org/10.3390/ijms232214257.
64. Sekhoacha M, Riet K, Motloung P, Gumenku L, Adegoke A, Mashele S. Prostate cancer review: genetics, diagnosis, treatment options, and alternative approaches. Molecules. 2022;27:5730. https://doi.org/10.3390/molecules27175730.
65. Bookstein R. Tumour suppressor genes in prostate cancer. In: Prostate cancer. Springer; 2001. p. 61–94. https://doi.org/10.1385/1-59259-009-8:61.
66. Tan BL, Norhaizan ME. Oxidative stress, diet and prostate cancer. World J Mens Health. 2021;39(2):195–207. https://doi.org/10.5534/wjmh.200014.
67. Pizzino G, Irrera N, Cucinotta M, Pallio G, Mannino F, Arcoraci V, Squadrito F, Altavilla D, Bitto A. Oxidative stress: harms and benefits for human health. Oxid Med Cell Longev. 2017;2017:8416763. https://doi.org/10.1155/2017/8416763. Epub 2017 Jul 27. PMID: 28819546; PMCID: PMC5551541.
68. Gonthier K, Poluri RTK, Audet-Walsh É. Functional genomic studies reveal the androgen receptor as a master regulator of cellular energy metabolism in prostate cancer. J Steroid Biochem Mol Biol. 2019;191:105367.
69. Bansal A, Simon MC. Glutathione metabolism in cancer progression and treatment resistance. J Cell Biol. 2018;217:2291–8.
70. Olinski R, Zastawny TH, Foksinski M, Barecki A, Dizdaroglu M. DNA base modifications and antioxidant enzyme activities in human benign prostatic hyperplasia. Free Radic Biol Med. 1995;18(4):807–13.
71. Doğru-Abbasoğlu S, Aykaç-Toker G, Koçak T, Unlüer E, Uysal M. Antioxidant enzyme activities and lipid peroxides in the plasma of patients with benign prostatic hyperplasia or prostate cancer are not predictive. J Cancer Res Clin Oncol. 1999;125(7):402–4.

72. Yossepowitch O, Pinchuk I, Gur U, Neumann A, Lichtenberg D, Baniel J. Advanced but not localized prostate cancer is associated with increased oxidative stress. J Urol. 2007;178(4):1238–44.
73. Thienger P, Rubin MA. Prostate cancer hijacks the microenvironment. Nat Cell Biol. 2021;23(1):3–5. https://doi.org/10.1038/s41556-020-00616-3.
74. Gevaert T, Van Eycke Y-R, Vanden Broeck T, Van Poppel H, Salmon I, Rorive S, et al. The potential of tumour microenvironment markers to stratify the risk of recurrence in prostate cancer patients. PLoS One. 2021;15(12):e0244663. https://doi.org/10.1371/journal.pone.0244663.
75. Chen YQ, et al. Testosterone ameliorates vascular aging via the Gas6/Axl signalling pathway. Aging. 2020;12(16):16111–25. https://doi.org/10.18632/aging.103584.
76. Bianchi-Frias D, Vakar-Lopez F, Coleman IM, Plymate SR, Reed MJ, Nelson PS. The effects of aging on the molecular and cellular composition of the prostate microenvironment. PLoS One. 2010;5(9):e12501. https://doi.org/10.1371/journal.pone.0012501. PMID: 20824135; PMCID: PMC2931699.
77. Morrison C, Thornhill J, Gaffney E. The connective tissue framework in the normal prostate, BPH and prostate cancer: analysis by scanning electron microscopy after cellular digestion. Urol Res. 2000;28(5):304–7. https://doi.org/10.1007/s002400000123.
78. Chen S, Zhu G, Yang Y, Wang F, Xiao Y-T, Zhang N, et al. Single-cell analysis reveals transcriptomic remodelling's in distinct cell types that contribute to human prostate cancer progression. Nat Cell Biol. 2021;23(1):87–98. https://doi.org/10.1038/s41556-020-00613-6.
79. Cunha GR, Donjacour AA, Sugimura Y. Stromal–epithelial interactions and heterogeneity of proliferative activity within the prostate. Biochem Cell Biol. 1986;64(6):608–14. https://doi.org/10.1139/o86-084.
80. Jennbacken K, et al. The prostatic environment suppresses growth of androgen-independent prostate cancer xenografts: an effect influenced by testosterone. Prostate. 2009;69(11):1164–75. https://doi.org/10.1002/pros.20965.
81. McKenzie S, Kyprianou N. Apoptosis evasion: the role of survival pathways in prostate cancer progression and therapeutic resistance. J Cell Biochem. 2006;97(1):18–32. https://doi.org/10.1002/jcb.20634.
82. Wong RS. Apoptosis in cancer: from pathogenesis to treatment. J Exp Clin Cancer Res. 2011;30(1):87. https://doi.org/10.1186/1756-9966-30-87.
83. Schneider P, Tschopp J. Apoptosis induced by death receptors. Pharm Acta Helv. 2000;74:281–6. https://doi.org/10.1016/S0031-6865(99)00038-2.
84. Karp G. Cell and molecular biology: concepts and experiments. 5th ed. New York: Wiley; 2008. p. 653–7.
85. Reed JC. Bcl-2 family proteins: regulators of apoptosis and chemoresistance in haematologic malignancies. Semin Haematol. 1997;34:9–19.
86. Hengartner MO. The biochemistry of apoptosis. Nature. 2000;407(6805):770–6. https://doi.org/10.1038/35037710.
87. McCarty MF. Targeting multiple signalling pathways as a strategy for managing prostate cancer: multifocal signal modulation therapy. Integr Cancer Ther. 2004;3(4):349–80. https://doi.org/10.1177/1534735404270757.
88. Navone NM, Troncoso P, Pisters LL, Goodrow TL, Palmer JL, Nichols WW, et al. p53 protein accumulation and gene mutation in the progression of human prostate carcinoma. J Natl Cancer Inst. 1993;85(20):1657–69. https://doi.org/10.1093/jnci/85.20.1657.
89. Datta K, Muders M, Zhang H, Tindall DJ. Mechanism of lymph node metastasis in prostate cancer. Future Oncol. 2010;6:823–36.
90. Body JJ, Casimiro S, Costa L. Targeting bone metastases in prostate cancer: improving clinical outcome. Nat Rev Urol. 2015;12:340–56.
91. Aceto N, Toner M, Maheswaran S, Haber DA. En route to metastasis: circulating tumour cell clusters and epithelial-to-mesenchymal transition. Trends Cancer. 2015;1:44–52.

92. Taichman RS, Cooper C, Keller ET, Pienta KJ, Taichman NS, McCauley LK. Use of the stromal cell-derived factor-1/CXCR4 pathway in prostate cancer metastasis to bone. Cancer Res. 2002;62:1832–7.
93. Jung Y, Wang J, Lee E, McGee S, Berry JE, Yumoto K, Dai J, Keller ET, Shiozawa Y, Taichman RS. Annexin 2–CXCL12 interactions regulate metastatic cell targeting and growth in the bone marrow. Mol Cancer Res. 2015;13:197–207.
94. Barthel SR, Hays DL, Yazawa EM, Opperman M, Walley KC, Nimrichter L, Burdick MM, Gillard BM, Moser MT, Pantel K, et al. Definition of molecular determinants of prostate cancer cell bone extravasation. Cancer Res. 2013;73:942–52.

Chapter 2
PSA: Myths and Reality

Rohit Deshpande and Ravimohan S. Mavuduru

PSA is the most significant tumour marker discovery of twentieth century

PSA is prostatic specific but not disease (cancer) specific

2.1 Introduction

Prostate-specific antigen (PSA) is a protein produced by the prostate gland, and its levels can be measured through a blood test. PSA testing is commonly used as a tool for early detection and monitoring of prostate cancer. However, there are several myths and misconceptions surrounding PSA, leading to confusion and misunderstanding. This write-up aims to shed light on the subject by exploring the truth behind these myths and the reality of PSA testing.

2.2 What is PSA?

Prostate-specific antigen (also known as hK3 i.e., human kallikrein 3), as the name suggests, is a protein that is produced exclusively by the prostate gland and is a member of the 'kallikrein' gene family. It plays a crucial role in the liquefaction of semen, aiding in sperm mobility and fertility. PSA is present in the blood in small amounts under normal circumstances. However, when there is an abnormality in the prostate, such as inflammation, infection, or cancer, PSA levels can become elevated.

PSA is normally bound to proteins on the blood and part of it is unbound or free. Thus several types of PSA exist, which are available for testing e.g., free PSA and total PSA and free/total PSA ratio etc. [1]. These are adjuncts, used when the PSA levels are elevated and the histological proof is inconclusive.

R. Deshpande · R. S. Mavuduru (✉)
Department of Urology, PGIMER Chandigarh, Chandigarh, India

N. Gupta (ed.), *A Clinical Guide for Cancer Prostate Patient*,
https://doi.org/10.1007/978-981-97-5357-4_2

2.3 Situations in Which PSA Gets Elevated

PSA is organ-specific rather than cancer-specific, it may be high in non-cancerous diseases including BPH and prostatitis. But even in symptomatic patients, PSA maintains its diagnostic significance for cancer detection. Hormonal therapies, even those used to treat benign disorders (often BPH), such as finasteride or dutasteride, have a negative impact on PSA values as well and can lead to falsely lowered PSA values [1].

1. Prostate cancer: One of the primary reasons for elevated PSA levels is prostate cancer. Cancerous cells within the prostate gland can cause an increase in PSA production and subsequently its release into the bloodstream, leading to higher levels detected in the blood, due to a distortion in the microscopic architectural pattern of the prostate gland (see Fig. 2.1) [2]. However, it is important to note that not all cases of elevated PSA indicate the presence of cancer. Other factors, such as age, prostate size, and overall health, should be taken into account before jumping to conclusions.

2. Prostate inflammation: Prostatitis, which refers to inflammation of the prostate gland, can also cause a rise in PSA levels. This condition may be a result of infection or other non-cancerous causes. It is crucial to differentiate between prostate inflammation and prostate cancer through additional diagnostic tests to avoid unnecessary alarm.
3. Benign Prostatic Hyperplasia (BPH): BPH is a non-cancerous enlargement of the prostate gland commonly experienced by aging men. The growth of the prostate can lead to increased PSA levels. While BPH is not cancerous, it can cause urinary symptoms that require medical attention.
4. Apart from this, procedures such as cycling, sexual activity, digital rectal examination, instrumentation etc. can also cause transient elevations in PSA (Fig. 2.2).

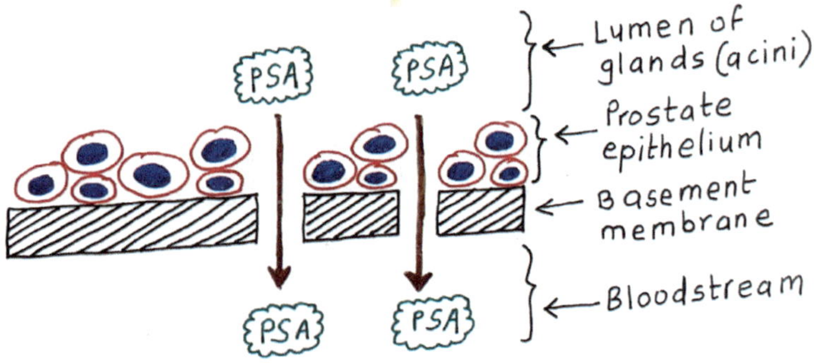

Fig. 2.1 Diagrammatic representation of loss of the tissue architecture in prostate cancer, causing increase of PSA in serum

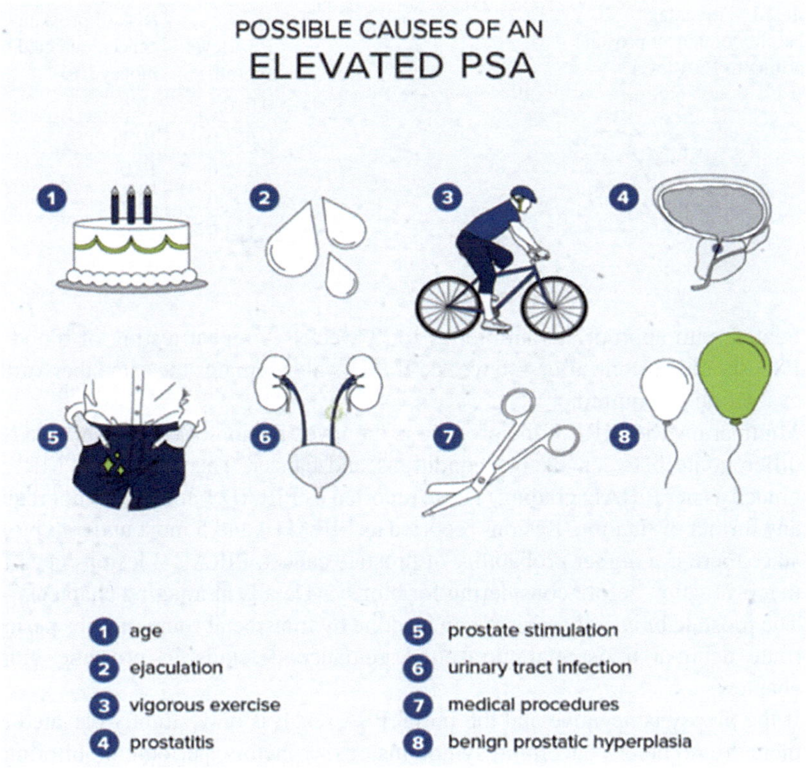

Fig. 2.2 Pictorial presentation of various causes of elevated PSA

2.4 What To Do If PSA Gets Elevated?

If PSA levels are elevated, it does not automatically indicate the presence of prostate cancer. PSA thresholds are not defined according to any accepted norms. Increased levels indicate an increased risk of prostate cancer; it is a continuous parameter. Many men may have prostate cancer despite having low serum PSA levels, making it impossible to set an ideal threshold for identifying clinically relevant but non-palpable prostate cancer [1] (Table 2.1).

Indian patients are believed to have a higher cut-off for prostate cancer. The proposed PSA cut-off in Indian men is 5.5 ng/dL [3]. For all practical purposes a PSA of >4 ng/mL, should get further evaluation. It is essential to consult with a healthcare professional to determine the underlying cause. Here, are a few steps to consider:

1. First step would be to rule out other causes of elevated PSA. A digital rectal examination is mandatory, which helps rule out prostatitis or infection of the prostate. If infection is suspected then a urine/semen culture must be done and

Table 2.1 Percentage probability of cancer prostate according to PSA level

PSA level (ng/mL)	Risk of prostate cancer, detected by biopsy (%)
0.0–0.5	6.6
0.6–1.0	10.1
1.1–2.0	17.0
2.1–3.0	23.9
3.1–4.0	26.9

treated with appropriate antibiotics for 3 weeks. A repeat testing of blood for PSA should be done after 4–6 weeks. If PSA value remains the same then further evaluation is required.

2. Multiparametric MRI of the prostate is the investigation of choice and can help differentiate between benign conditions and cancer. The lesions on MRI are graded as per PIRAD grading. Those reported as PIRAD 1 and 2 does not require any further evaluation. Lesions reported as PIRAD 4 and 5 must undergo biopsy since there is a higher probability of prostate cancer. PIRAD 3 lesions need further evaluation before considering for biopsy. (Details in imaging chapter).
3. The prostate biopsy if required can be done by transrectal route or trans perineal route using a transrectal ultrasound guidance. (Details in prostatic biopsy chapter).
4. If the biopsy is negative and the initial PSA result is only slightly elevated and there are no other concerning symptoms or risk factors, periodic monitoring of PSA is required. PSA monitoring after therapy refers to the regular measurement of PSA levels in a patient who has undergone treatment for prostate cancer. This approach allows for observation over time to detect any significant changes.
5. Along with PSA, other parameters such as PSA doubling time, PSA density, proportion of free PSA, rate of rise in PSA, PHI (Prostate Health index) etc. are performed in follow-up as an adjunct to PSA level for further evaluations.
6. Typically, PSA testing is recommended once in 3 months. If PSA values continue to rise or the suspicion of underlying prostate cancer is high then a repeat biopsy with increased number of cores up to 18–24 cores is advisable or targeted biopsy using MRI fusion technology is recommended. Subsequent to this if the biopsy is negative for malignancy, then a periodic follow-up should be done.
7. Although its role in screening is controversial, PSA has a major role in monitoring of the treatment, once the prostate cancer is diagnosed.

2.5 Importance of PSA in Prostate Cancer Management

1. **Assess treatment response**: PSA levels can be used to evaluate how well the therapy has worked in reducing or eliminating cancerous cells. A decrease in PSA levels after treatment is generally a positive sign, indicating a favourable response to therapy.
2. **Detect recurrence**: Rising PSA levels after treatment may indicate cancer recurrence. If PSA levels start to increase after a period of stability or decline, it could be a sign that cancer cells have returned, and further investigation and management may be necessary.
3. **Early detection of metastasis**: PSA monitoring can help in detecting potential spread of cancer to other parts of the body (metastasis). If PSA levels rise significantly and consistently, it may suggest that the cancer has spread beyond the prostate.
4. **Guide treatment decisions**: PSA trends can provide valuable information for making decisions about further treatment options. For instance, a slow and gradual rise in PSA levels may indicate the need for close monitoring or the initiation of salvage therapies, while a rapid and steep increase might prompt more aggressive interventions.

It's important to note that PSA monitoring is just one aspect of post-therapy follow-up for prostate cancer patients. Clinical assessments, imaging tests, and other factors are also taken into consideration to provide a comprehensive evaluation of a patient's condition.

The frequency of PSA monitoring and the specific PSA level thresholds that trigger additional actions may vary depending on the individual patient's characteristics and the type of therapy they received. These decisions are usually made by the patient's healthcare team based on guidelines and their clinical expertise. If you or someone you know is undergoing or has completed prostate cancer treatment, it's essential to discuss the post-therapy monitoring plan with the healthcare provider to ensure appropriate follow-up and care.

2.6 When to Get PSA Done?

The decision of when to get a PSA test should be based on various factors, including age, family history, and overall health. There are differing recommendations among medical organizations regarding the timing and frequency of PSA screening. Here are some general guidelines:

1. Age: The American Urological Association (AUA) suggests that men should consider discussing PSA testing with their doctor starting at age 55. However, for individuals at higher risk, such as those with a family history of prostate cancer or African American men, the discussion may begin at age 40–45. It is not

recommended to perform PSA testing in men >70 years of age and with a life expectancy of <10–15 years.
2. Shared Decision-Making: It is important to engage in a conversation with your doctor about the benefits and risks of PSA screening. Together, you can assess your individual risk factors, discuss your preferences, and make an informed decision about when to initiate PSA testing. Before getting PSA testing, it is important to understand that if PS is high then further investigations and treatment will be required if diagnosis is confirmed.
3. Frequency: The frequency of PSA testing depends on individual risk factors and the initial PSA level. If PSA level is low (less than 2 ng/mL) and there are no concerning symptoms or risk factors, recommend testing is every 2 years. However, if your PSA level is higher or if there are other risk factors present, more frequent testing may be necessary.
4. Screening: Screening of the population with PSA testing is not recommended. However, opportunistic screening is advised in patient above the age of 40 years who present with lower urinary tract symptoms (LUTS). This helps in further evaluation and management.

2.7 Conclusion

PSA testing is a valuable tool in the early detection and monitoring of prostate cancer. However, it is important to understand the myths and realities surrounding PSA levels. Elevated PSA levels can indicate various conditions, including prostate cancer, inflammation, or benign enlargement of the prostate. It is crucial to consult with a healthcare professional to determine the cause of elevated PSA and develop an appropriate management plan. Additionally, the decision of when to initiate PSA testing should be based on individual risk factors and shared decision-making between the patient and the healthcare provider. By demystifying PSA and promoting informed discussions, we can ensure that individuals receive appropriate care and accurate information regarding their prostate health.

References

1. EAU guidelines on prostate cancer, 2023.
2. Partin AW, et al. Campbell-Walsh textbook of urology. 12th ed. Elsevier; 2020.
3. Kumar A, et al. The Urological Society of India guidelines for the evaluation and management of prostate cancer. Indian J Urol. 2022;38(4):252–7.

Chapter 3
Diagnosis and Clinical Staging of Prostate Cancer

Brusabhanu Nayak

Proper Diagnosis, staging and risk stratification is very important for management of Prostate Cancer

- Prostate cancer is mostly asymptomatic at an early stage. The presence of symptoms suggests advanced disease. In western countries, most prostate cancers are diagnosed at an early stage due to wide spread screening with serum PSA [1]. Hence, locally advanced and metastatic disease are less common presentation.
- In India, majority of patients are diagnosed at an advanced stage after becoming clinically symptomatic. However, in urban India patients are also diagnosed at an early stage due to increase in PSA screening.
- After diagnosis, proper staging of the prostate cancer is done to determine the extent of disease and further management. Since, the management of localized cancers, locally advanced/metastatic cancers are different, proper staging is extremely important.
- Prostate cancer can be "clinically significant" or "clinically insignificant". The distinction between these types is important because clinically significant cancers can lead to morbidity and mortality whereas "clinically insignificant cancers" don't cause any harm to the patient. Unless this distinction is clearly made, these insignificant cancers can have higher risk of being over treated and sometimes the treatment itself cause more harmful side effects (Table 3.1).
- **Diagnosis**: Diagnosis of prostate cancer is done by screening tests followed by confirmatory tests. Screening tests include serum PSA and digital rectal examination (DRE). After screening, diagnosis is confirmed by needle biopsy of the prostate gland.

B. Nayak (✉)
Department of Urology, All India Institute of Medical Sciences, New Delhi, India

N. Gupta (ed.), *A Clinical Guide for Cancer Prostate Patient*,
https://doi.org/10.1007/978-981-97-5357-4_3

Table 3.1 Recommendations for screening and early detection of prostate cancer as per European guidelines [2]

Recommendations for screening and early detection	Strength rating
• Do not subject men to prostate-specific antigen (PSA) testing without counselling them on the potential risks and benefits	Strong
• Offer an individualised risk-adapted strategy for early detection to a well-informed man with a life-expectancy of at least 10–15 years	Weak
• Offer early PSA testing to well-informed men at elevated risk of having PCa: • Men from 50 years of age • Men from 45 years of age and a family history of PCa • Men of African descent from 45 years of age • Men carrying BRCA2 mutations from 40 years of age	Strong
• Offer a risk-adapted strategy (based on initial PSA level), with follow-up intervals of 2 years for those initially at risk: • Men with a PSA level of >1 ng/mL at 40 years of age • Men with a PSA level of >2 ng/mL at 60 years of age • Postpone follow-up to 8 years in those not at risk	Weak
• In asymptomatic men with a prostate specific antigen (PSA) level between 3–10 ng/mL and a normal digital rectal examination (DRE), repeat the PSA test prior to further investigations	Weak
• In asymptomatic men with a PSA level between 3–10 ng/mL and a normal DRE, use one of the following tools for biopsy indication: • Risk-calculator, provided it is correctly calibrated to the population prevalence • Magnetic resonance imaging of the prostate	Strong
• An additional serum, urine biomarker test	Weak
• Stop early diagnosis of PCa based on life expectancy and performance status; men who have a life-expectancy of <15 years are unlikely to benefit	Strong

- **Digital rectal examination (DRE)**: DRE is commonly used to assess the prostate gland. Before the availability of PSA testing, clinicians used to rely on DRE for detection of prostate cancer. DRE in the hands of experienced clinicians has better reproducibility. However, only DRE misses a significant proportion of prostate cancer. The value of DRE screening at serum PSA <4 ng/mL is limited. PSA testing along with DRE improves the detection of prostate cancer. DRE and serum PSA testing are complimentary to each other in the diagnosis of prostate cancer. In some patients, with prostate cancer PSA can be normal with abnormal DRE findings and in some others, PSA can be high with normal DRE findings. Hence, it is recommended to do both DRE and serum PSA as screening tests for prostate cancer diagnosis [3].
- **Serum PSA**: PSA is a member of the human Kallikrein family. It is secreted in high concentrations in seminal fluid. PSA levels in the serum vary with age, race, prostate volume. Normal serum PSA range is 0–4 ng/mL. PSA is organ specific but not cancer specific. PSA can be elevated in chronic prostatitis, acute urinary retention and urinary tract infection. Elevated serum PSA in these situations should be considered with caution and the PSA level should be reconfirmed after

the treatment of above conditions. The details on PSA have been discussed in the previous chapter.

- After the screening tests, further diagnostic evaluation is performed by either multiparametric MRI (MP-MRI) prostate followed by MR-fusion biopsy or with standard TRUS biopsy depending upon the DRE finding and MP-MRI findings. Following scenarios can help us better understand the diagnostic pathway for prostate cancer.
- ↑ Serum PSA with diffuse hard prostate → MP MRI → Standard TRUS biopsy
- Normal Serum PSA with diffuse hard prostate → MP MRI → Standard TRUS biopsy
- ↑ Serum PSA with Normal DRE → MP MRI prostate

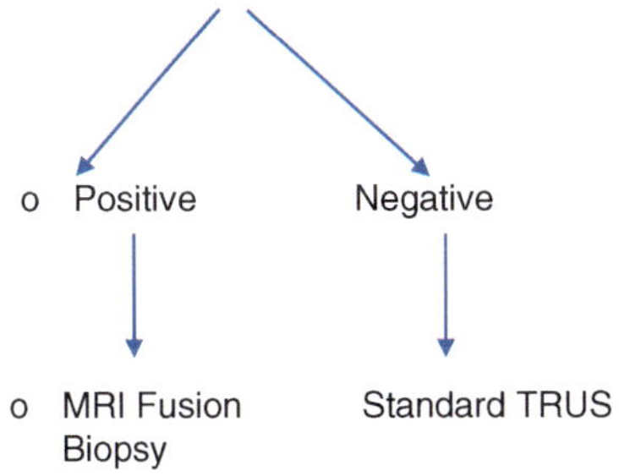

- Details on multi parametric MRI prostate and biopsy have been discussed in separate chapters.
- **Clinical staging**: Once the diagnosis of the prostate cancer is confirmed, the next step is to do the staging of the disease to see for the extent of involvement. The decision to do further staging evaluation is guided by the treatment options available for the patient, taking into consideration the life expectancy of the patient.
- Clinical staging is different from pathological staging. Clinical staging is the assessment of disease extent with pretreatment parameters and pathological staging is done after the prostate removal and when the final histopathology is available.
- Clinical staging is done both for the assessment of the prognosis and to decide regarding appropriate treatment. Clinical staging in prostate cancer is performed using parameters like DRE, serum PSA, TRUS biopsy findings and radiological imaging.
- After the diagnosis is confirmed with needle biopsy of the prostate further imaging are performed to confirm the extent of spread of the disease. Additional radiological imaging is performed to know if the disease is localized to prostate or metastasized to other distance organs. Prostate cancer can spread to distant non regional lymph nodes in the retroperitoneum, bones, other visceral organs like liver and lungs.
- After confirmation of the diagnosis, the prostate cancer patients are risk stratified into various risk categories because the management of prostate cancer is decided on the basis of risk categories. The risk stratifications for prostate cancer are

divided into very low risk, low risk, intermediate risk, high risk and very high-risk categories (Tables 3.2 and 3.3).

- **Imaging**: Various imaging modalities are available and has been used for staging prostate cancer. Different imaging modalities that are commonly used in the staging include whole body bone scan, CECT abdomen and pelvis, MRI abdomen and pelvis and PSMA PET scan etc.
- After confirmation of the prostate cancer diagnosis, accurate staging of the disease is performed using the above-mentioned imaging modalities. These imaging modalities can be used as single investigation or combination of more than one modality depending upon the clinical need. The type of imaging modalities

Table 3.2 NCCN guidelines risk stratifications

Risk group	Clinical/pathological feature
• Very low	• T1c AND • Grade group 1 AND • PSA <10 ng/mL AND • Fewer than 3 prostate biopsy fragments/cores positive, ≤50% cancer in each fragment/core AND • PSA density <0.15 ng/mL/g
• Low	• T1–T2a AND • Grade group 1 AND • PSA <10 ng/mL AND • Does not qualify for very low risk
• Intermediate (favorable)	• No high or very high risk features • No more than one intermediate risk factor: – T2b–T2c OR – Grade group 2 or 3 – PSA 10–20 ng/mL • Grade group 1 or 2 • Percentage of positive biopsy cores <50%
• Intermediate (unfavorable)	• No high or very high risk features • Two or three of the intermediate risk factors: – T2b–T2c – Grade group 2 or 3 – PSA 10–20 ng/mL • AND/OR • Grade group 3 • AND/OR • ≥50% of positive biopsy cores
• High risk	• No very high risk features • T3a OR • Grade group 4 or 5 OR • PSA >20 ng/mL
• Very high risk	• T3b–T4 • Primary Gleason pattern 5 OR • Two or three high-risk features OR • >4 cores with Grade group 4 or 5

Table 3.3 TNM staging for prostate cancer [4]

Category	Criteria
Clinical (cT) T stage	
• TX	• Primary tumor cannot be assessed
• T0	• No evidence of primary tumor
• T1	• **Clinically inapparent tumor that is not palpable**
• T1a	• Tumor incidental histologic finding in 5% or less of tissue resected
• T1b	• Tumor incidental histologic finding in more than 5% of tissue resected
• T1c	• Tumor identified by needle biopsy found in one or both sides, but not palpable
• T2	• **Tumor is palpable and confined within prostate**
• T2a	• Tumor involves one-half of one side or less
• T2b	• Tumor involves more than one-half of one side but not both sides
• T2c	• Tumor involves both sides
• T3	• Extra prostatic tumor that is not fixed or does not invade adjacent structures
• T3a	• Extra prostatic extension (unilateral or bilateral)
• T3b	• Tumor invades seminal vesicle(s)
• T4	• Tumor is fixed or invades adjacent structures other than seminal vesicles, such as external sphincter, rectum, bladder, levator muscles, and/or pelvic wall
N category	
• Nx	• Regional lymph nodes were not assessed
• N0	• No positive regional lymph nodes
• N1	• Metastases in regional lymph node(s)
M category	
• M0	• No distant metastasis
• M1	• Distant metastasis
• M1a	• Nonregional lymph node(s)
• M1b	• Bone(s)
• M1c	• Other visceral sites with or without bone disease

recommended for staging of the prostate cancer is decided on the basis of risk stratification of the prostate cancer.

- **Bone scans**: According to the recommendation of various guidelines, bone scans are performed in following situations [5, 6]. (1) Patients with clinical stage T1 and a serum PSA level greater than 20 ng/mL. (2) Clinical stage T2 and serum PSA greater than 10 ng/mL. (3) Gleason score 8–10 (grade group 4–5). (4) Clinical stage T3 or T4. (5) Patients with clinical symptoms like bone pain. Routine bone scans are not necessary in low risk and very low risk prostate cancer patients [7].
- **CECT scan of abdomen and pelvis, MRI abdomen and pelvis**: Cross sectional imaging like CECT abdomen with pelvis or MRI is used for locoregional staging and evaluation of regional lymph nodes.
- **PSMA PET/CT scan**: PSMA PET/CT scan and whole-body MRI provide more sensitive detection of lymph node and bone metastases than conventional work

Table 3.4 Recommendations for prostate cancer staging as per the European Urology guidelines is as follows

Recommendations	Strength ratings
Any risk grouping	
• Use pre-biopsy MRI for local staging information	• Weak
• Treatment should not be changed based on PSMA PET/CT findings, in view of current available data	• Strong
Low-risk localised disease	
• Do not use additional imaging for staging purposes	• Strong
Intermediate-risk disease	
• In ISUP grade 3, include at least cross-sectional abdominopelvic imaging and a bone-scan for metastatic screening	• Weak
High-risk localised disease/locally-advanced disease	
• Perform metastatic screening including at least cross-sectional abdominopelvic imaging and a bone-scan	• Strong
• When using PSMA-PET or whole body MRI to increase sensitivity, be aware of the lack of outcome data of subsequent treatment changes	• Strong

up with bone scan and abdominopelvic CT scan. In view of the of results from multi center proPSMA randomized controlled trial PSMA PET/CT may be considered as single more sensitive imaging modalities for staging of the high-risk prostate cancer patients [8, 9]. However, the prognosis and ideal management of patients diagnosed as metastatic by these more sensitive imaging modalities are unknown. Hence, in absence of studies showing therapeutic survival benefits cautions should be used while taking therapeutic decisions on the basis these imaging modalities (Table 3.4).

References

1. Loeb S, et al. Pathological characteristics of prostate cancer detected through prostate specific antigen based screening. J Urol. 2006;175:902.
2. Boyle HJ, et al. Updated recommendations of the International Society of Geriatric Oncology on prostate cancer management in older patients. Eur J Cancer. 2019;116:116. https://pubmed.ncbi.nlm.nih.gov/31195356/
3. Naji L, et al. Digital rectal examination for prostate cancer screening in primary care: a systematic review and meta-analysis. Ann Fam Med. 2018;16:149.
4. Abuzallouf S, et al. Baseline staging of newly diagnosed prostate cancer: a summary of the literature. J Urol. 2004;171:2122.
5. Briganti A, et al. When to perform bone scan in patients with newly diagnosed prostate cancer: external validation of the currently available guidelines and proposal of a novel risk stratification tool. Eur Urol. 2010;57:551.
6. Lin Y, et al. When to perform bone scintigraphy in patients with newly diagnosed prostate cancer? A retrospective study. BMC Urol. 2017;17:41.
7. O'Sullivan JM, et al. Broadening the criteria for avoiding staging bone scans in prostate cancer: a retrospective study of patients at the Royal Marsden Hospital. BJU Int. 2003;92:685.

8. Hofman MS, et al. Prostate-specific membrane antigen PET-CT in patients with high-risk prostate cancer before curative-intent surgery or radiotherapy (proPSMA): a prospective, randomised, multicentre study. Lancet. 2020;395:1208.
9. Cornford P, et al. Prostate-specific membrane antigen positron emission tomography scans before curative treatment: ready for prime time? Eur Urol. 2020;78:e125. https://pubmed.ncbi.nlm.nih.gov/32624287/.

Chapter 4
Imaging in Prostate Carcinoma

Kulbir Ahlawat and Aprajita Singh

Imaging is important to define anatomy of the prostate, diagnose various diseases and helps in planning and management. There are several modalities for imaging for the prostate gland. as per the patient characteristics and PSA Level, type of imaging decided.

4.1 Introduction

- Prostate is a part of male reproductive system that serves dual functions, both as an unpaired accessory gland as well as a muscular mechanical switch between urination and ejaculation. Its glandular function includes production of fluid (containing glycoprotein, prostate specific antigen) that forms part of semen that help neutralize the acidity of the vaginal tract prolonging lifespan of the sperm.
- Prostate cancer is one of the top ten leading cancers in India usually affecting men in the age group of 65 years and above. However, according to recent studies have reported an increase in incidences of prostate cancers in younger age groups (35–44 and 55–64 years), more in metropolitan cities.
- The most common form of prostate cancer is adenocarcinoma that mostly occurs in the peripheral zone. It usually develops asymptomatically and in the initial stages may resemble findings of benign prostatic hypertrophy, however in later stages may show extra-prostatic extension into surrounding areas with distant metastases to bones and central nervous system.
- Pathologists traditionally used Gleason score to grade the prostate cancer. It comprises of two grades, one of the most predominant pattern (primary) and the other of the second most predominant pattern. In case there is a worse grade than the primary pattern, it is considered as the second pattern even though it could be the third most prevalent.

K. Ahlawat (✉) · A. Singh
Department of Radiology, Medanta, The Medicity, Gurugram, India
e-mail: kulbir.ahlawat@medanta.org

N. Gupta (ed.), *A Clinical Guide for Cancer Prostate Patient*,
https://doi.org/10.1007/978-981-97-5357-4_4

4.2 Basic Anatomy to See in Imaging

- Prostate gland in adults is a funnel/walnut shaped exocrine gland situated between bladder anteriorly and rectum posteriorly surrounded by a pseudo fibromuscular capsule (which most likely represents a thin inseparable connective tissue rather than a true capsule). Glandular and fibromuscular tissue mainly form the prostate gland that surrounds proximal male urethra which traverses through the central portion of the gland.
- Base of the prostate gland which constitutes the largest part of the gland usually refers to the cephalic portion of the gland which is continuous with the bladder neck. The muscle bands from the wall of bladder penetrate the fibromuscular capsule of the gland thereby establishing a connection between base of prostate gland and bladder neck. The prostatic urethra enters the gland at base of the gland. The midland included verumontanum in the mid prostatic urethra and refers to the middle one-third of the gland. The apex refers to the caudal end that is adherent to urogenital diaphragm/perineal membrane (consisting of deep transverse perineal muscle, external urethral sphincter and compressor urethrae muscles separating superficial perineal pouch from upper pelvis). The apex lies in close proximation with bulbourethral glands and bulb of the penis.
- Seminal vesicles lies posterior to prostate gland in close proximity. The ducts of seminal vesicles join the vas deferens ampullae at the prostate base to form the ejaculatory duct which then extends within the central portion of prostate in the midline. The ejaculatory duct joins the urethra at the level of a longitudinal ridge on the posterior wall of prostatic urethra also called as verumontanum, also called as the seminal colliculus (usually located near the apex at midpoint of prostatic urethra flanked on either side by opening of ejaculatory ducts). A small epithelium lined diverticulum may be seen at apex of verumontanum called utricle cyst which represents fatal remnant of urogenital sinus and is homologous to female uterus.
- Arterial supply of prostate is from internal iliac arteries via prostate-vesical branches. The prostatic arteries usually arise from inferior vesicle, internal pudendal arteries and to a lesser extent the inferior rectal artery. These arteries usually course along medial and inferior surface of prostate along veins and nerves of prostatic plexus forming neuromuscular bundle. The prostatic and inferior vesical arteries further branching into urethral and capsular arteries. The urethral arteries follow the course of prostatic urethra supplying one-third of the glandular tissue while the capsular arteries supply the remaining two-thirds.
- The venous drainage of the prostate gland which is mainly constituted by prostatic venous plexus situated between the capsule and fascial layer. These small veins join the veins of penis and urinary bladder to ultimately drain into internal iliac veins. The internal iliac veins are also known to connect with vertebral venous plexus, which is considered to be the route of bone metastasis which are commonly seen in prostate carcinoma [1] (Fig. 4.1).

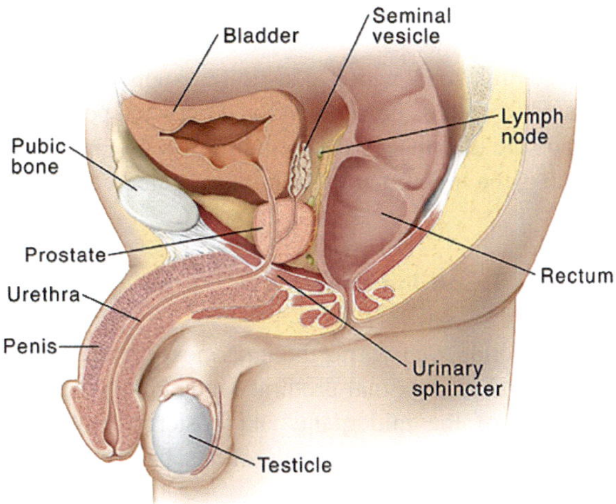

Fig. 4.1 Anatomy of the prostate gland

4.2.1 Zones of Prostate

- The main anatomical landmark of prostate is the urethra which also divides the gland into two-parts, namely the anterior fibromuscular part and the posterior glandular portion. The posterior glandular tissue accounts for two-thirds of the gland and is primarily delineated into four zones.

4.2.1.1 Peripheral Zone

The peripheral zone consists of approximately 70% of the glandular tissue along most of the apex of the gland, lateral and posterior aspects of the mid gland. Its duct drains into distal urethral segment between the apex and verumontanum. Sono graphically, it shows homogenous isoechoic echotexture separated from the central zone by a thin hyper echoic band called the surgical capsule (commonly seen a semicircular shaped band which borders the central zone laterally and posteriorly). On MR T2W images, the peripheral zone is shows homogeneously hyperintense signal. Posteriorly, it is separated from rectum by a thin capsule and abuts neurovascular bundles bilaterally which course anteromedially entering prostate at 5 and 7 o'clock. Most of the cases of prostate cancers occur in peripheral zone (accounting for ~70% of cases).

4.2.1.2 Central Zone

The central zone which constitutes approximately 25% of the glandular tissue located at base of prostate wedged between peripheral and transition zones. It surrounds the ejaculatory ducts and narrows to an apex at verumontanum. Prostatitis, which is the inflammation of prostate gland commonly affects the central zone.

4.2.1.3 Transition Zone

The transition zone forms the innermost part of the gland containing two small lobules lateral to proximal urethral segment superior to verumontanum i.e. anterolateral part of the mid gland and posterior part of base. Anteriorly it abuts the fibromuscular stroma, central zone at base posteriorly and peripheral zone at the mid gland and apex.

As the age progresses, the transition zone being the most common site for benign prostatic hyperplasia starts accounting for a much greater percentage of glandular tissue.

The central and transition zones of prostate cannot be differentiated on MRI as they show similar signal intensity, hence they are commonly referred to as central gland.

4.2.1.4 Anterior Fibro Muscular Stroma

This represents non-glandular tissue comprising primarily of smooth muscle which is contiguous with detrusor vesicae. It constitutes the anterolateral border of prostate and is connected to the inferior edge of pubic symphysis by puboprostatic ligament. On MR it shows T2 hypo intensity as it is devoid of glandular tissue (Fig. 4.2).

4.3 Surrounding Structures

Anterior to prostate lies pubic symphysis and retropubic space (cave of Retzius). Puboprostatic ligaments connect posterior aspect of pubic symphysis to outer longitudinal muscle of bladder adjacent to anterior margin of internal sphincter creating a ‘detrusor apron’.

Posterior to prostate lie rectal ampulla, denonvillier’s fascia along with prostatic venous and nervous plexus.

Laterally on both sides are the elevator ani muscles, pelvic fascia (parietal and visceral layers) and neuromuscular triangles.

Superiorly the neck of urinary bladder and internal urethral sphincter are located and inferiorly, the urogenital diaphragm and external urethral sphincter are located.

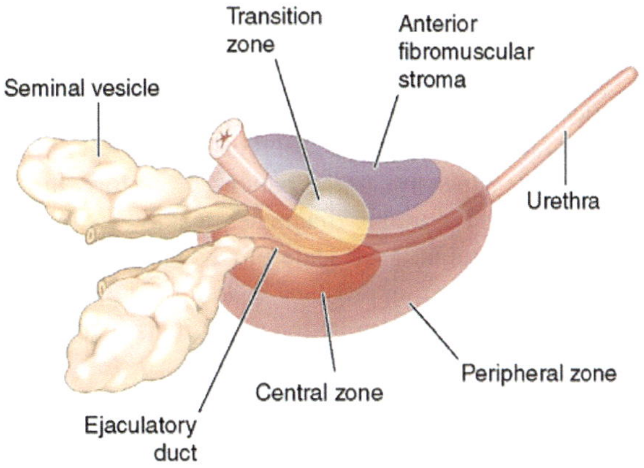

Fig. 4.2 Zonal anatomy of the prostate

The endopelvic fascia fuses with anterior fibromuscular stroma of prostate, levator ani muscles and pelvic side walls, covering rectum, urinary bladder and prostate. The periprostatic fascia covers lateral surfaces of prostate and consists of multiple layers made of mixed collagenous and adipose tissue. The levator ani fascia lying outer to prostatic fascia merge into periprostatic fascia and dorsally turns into pararectal fascia which segregates rectum from the elevator ani muscle. The prostatic fascia covers the prostate capsule and extends antero-medially to neuromuscular bundle. The posterior surface of prostate and seminal vesicles are covered by denonvillier's fascia (recto-prostatic fascia). These fascia's represent surgical dissection planes for radical prostatectomy that can be recognizable on MRI.

The neuromuscular bundle lie laterally and along the postern-lateral angle of the prostate and seminal vesicles as a bundle. They can also appear spread out within the adipose and connective tissue between prostate and elevator ani.

Posterior to the central zone lie seminal vesicles which also adjoin posterior wall of bladder. They are seen as multi cystic structures appearing hyper intense on T2W images and lower signal compared to periprostatic fat on T1W images. Medial to the vesicles lie the ampulla of vas deferens which are seen as thickwalled tubular structures. The ejaculatory ducts form at the junction of the vas defers ampulla and the seminal vesicles.

4.4 Ultrasound Imaging in Prostate Cancer

The prostate gland is most commonly affected by diseases such as prostatitis, benign prostatic hypertrophy and prostate cancer. Prostate cancer screening consists of serum levels of prostate specific antigen (PSA) along with digital-rectal examination (DRE) and imaging studies. The most common indications for prostate sonographic examinations include presence of palpable lesions, raised PSA levels, suspicion of prostatic inflammation or to assess the clinical symptoms related to urination and/or ejaculation.

In recent years, multi-parametric MRI (mp MRI) is considered to be the gold standard in diagnosing and staging prostate cancer, however it has certain limitations in the form of long duration of study, difficulty in interpreting imaging results and is more expensive to the patient. Therefore, ultrasound assessment of prostate remains the initial tool to diagnose prostatic lesions.

Trans abdominal grey scale ultrasound is regarded to show low sensitivity and specificity in detecting prostatic lesions owing to lower resolutions of low frequency probes required to penetrate the overlying structures to reach the prostate, therefore only gross abnormalities can be diagnosed. It is however useful in assessing prostatic volume. The volume of prostate gland on ultrasound is assessed in three dimensions and is measured using the following formula:

Prostate volume = Length × Breadth × Height × 0.5236.

Normal volume of prostate in young measures less than 25 cc. Prostate size grading is done on the basis of the calculated prostate volume: Grade I (26–40 cc), Grade II (41–60 cc) and grade III (>61–80 cc), Grade IV (>80 cc) [2] (Fig. 4.3).

4.4.1 Transrectal Ultrasound

Transrectal ultrasound uses a high frequency probe (usually 9 12 MHz). Prior information obtained from DRE and PSA should be available. The patient is asked to empty the bladder in effort to reduce the discomfort. The patient is then placed in left lateral decubitus position with one knee flexed towards the chest mimicking a fetal position. Per-rectal examination and lubrication using lignocaine gel is performed to reduce the pain following which the transducer with a probe cover is inserted in the rectal cavity.

The normal prostate gland is seen as a crescent shaped structure at the base of bladder appearing symmetrical in contour and surrounding the prostatic urethra.

Images of the base, mid gland and apex region are acquired in transverse plane. Superior to the level of verumontanum is the base of prostate showing homogenous echo texture and appearing isoechoic to surrounding tissue. A linear hypo echoic line, arcuate line can sometimes be visualized in the central portion of base which predominantly comprises of central gland and separates it from adjacent peripheral gland. As the transducer is moved inferiorly, the mid gland is visualized at which level urethra passes and more of the central gland can be visualized which appears relatively hyper echoic in comparison to peripheral gland. It is also at this level where the size of the prostate gland is measured in transverse and antero-posterior dimensions. As the age of the patient progresses, the prostate gland may change its shape to an ovoid structure owing to the bulge of the central gland anteriorly due to benign prostatic hyperplasia. Further inferiorly apex of the gland can be visualized which is usually more circular and shows relatively heterogenous echo texture than

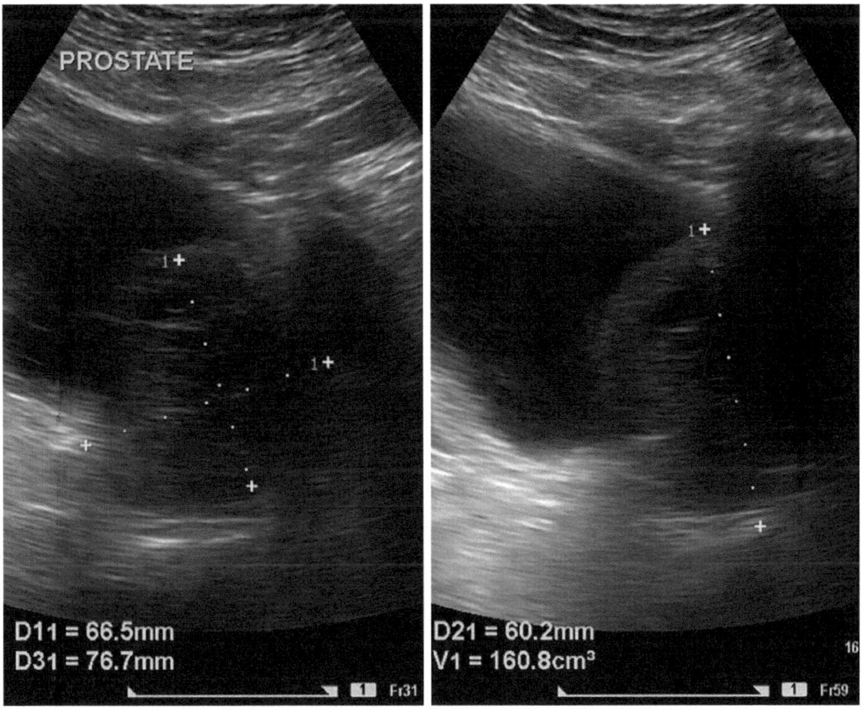

Fig. 4.3 Tans abdominal ultrasound (source—personal)

the homogenous base. Since 70% of the incidence of prostate cancers occur in peripheral zone, this is a common location for the malignant lesions.

For a complete prostatic evaluation, the seminal vesicles should also be visualized which lie lateral and superior to the base of prostate and normally appear hypoechoic. In cases, where the seminal vesicles appear enlarged or show heterogenous echo texture further investigations should be carried to rule out any obstructive process. The vas deferens appear thick-walled structures entering the base of bladder and are usually visualized between the seminal vesicles. In transverse sections they appear as two donut-shaped structures and show tubular orientation on sagittal images. Normally they should not contain any fluid within. Presence of fluid should indicate an obstructive process and should also be investigated further.

Prostatic lesions (benign or malignant) show different sonographic patterns. Prostatic carcinoma in most cases can be seen as a hypo echoic nodule compared to surrounding gland and occurs mostly in the peripheral zone at the level of apex. However, in many cases they may appear isoechoic or hypoechoic thereby causing difficulty in their diagnosis. Also the transitional zone shows more and more heterogeneity of echo texture as the age progresses, it makes the distinction between the glandular tissue and cancer even more difficult. Due to these reasons, TRUS has a low sensitivity and specificity.

There are few collateral or associated findings however that may aid in diagnosing malignant nature of a lesion. These malignant characteristics include irregularity of echo texture and asymmetrical enlargement of the involved zone of prostate, capsular bulge or extension of the lesion's contact surface beyond the prostate capsule around 2–3 mm which likely suggests presence of extra capsular extension.

Uncontrolled proliferation of blood vessels seen within neoplastic tissue is a typical feature in carcinomas known as angiogenesis. Therefore, the application of color Doppler and power Doppler help to demonstrate this phenomenon in areas of suspected focal neoplastic proliferation with power Doppler being the most sensitive tool to assess the flow in the blood vessels. There are different flow patterns of color flow that have been demonstrated in color Doppler, namely diffuse (most common), focal or surrounding (Fig. 4.4).

4.5 Recent Advances in Sonographic Techniques

Contrast enhanced ultrasound (CEUS) is based on administering gas filled micro-bubbles (measuring less than red blood cells) which are encapsulated in surfactant or lipid based envelope. These owing to their size can reach the capillary bed, thereby assessing the tissue micro-perfusion. Also since the ultrasound contrast agents are blood pool agents, it reaches the vascular areas without spreading into adjacent interstices of surrounding tissue which may obscure the visualization of the micro-vascularization of the nodule. The study is usually performed using a transrectal probe with the help of dedicated post-contrast settings which have low acoustic power and mechanical index that allow the detection of micro-bubbles. Once the suspicious nodule is identified, it is followed by administration of contrast agent during which the probe is focused on the suspected nodule. A malignant prostatic nodule is characterized by greater density of micro-vessels which shows irregularity in appearance as well as distribution in comparison to normal adjacent prostate gland which usually shows regular linear orientation of vessels from neuromuscular bundle which enters the glandular parenchyma.

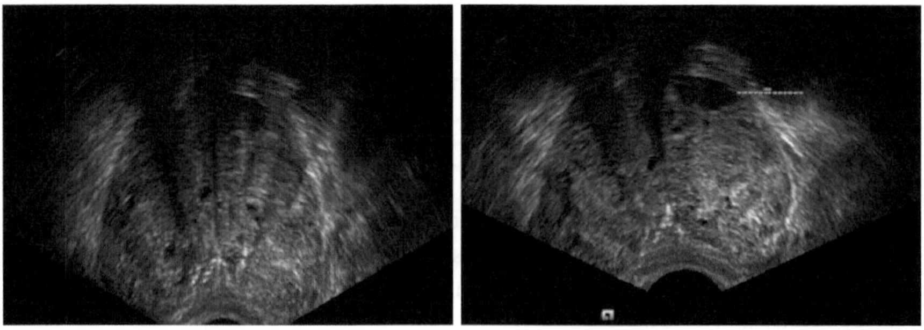

Fig. 4.4 Transrectal ultrasound showing suspicious lesion in apical region (source—self)

CEUS also helps in quantitative assessment of blood flow within the nodule by using a software that derives time-intensity curves (TIC) that includes parameters such as arrival time (AT), mean transit time (MTT), time to peak (TTP) and peak intensity (PI). These are evaluated by placing ROI in the region of interest and on any portion of healthy gland for comparison which show high and rapid enhancement peaks with duration of less than 20 s from time of administration. In order to remove operator dependance variability, software are available where the enhancement curves of the entire prostate are made and the area showing abnormal enhancement is highlighted. This is called "Parametric Dispersion CEUS".

Elastography: neoplastic tissues undergo a series of changes that alter their architecture such as micro-vascularization, increase in cellular density, collagen deposition and stromal reaction in surrounding periprostatic tissue. These lead to greater rigidity of neoplastic tissue in comparison to adjacent healthy gland which can be assessed by using elastography. It not only characterizes the degree of elasticity or stiffness of a suspected nodule but also helps identify areas that are not visible on conventional B-mode TRUS. Elastography is usually performed after measurement of prostatic volume, proper identification of suspected nodule and careful assessment of surrounding tissue. There are two types of elastography performed, namely strain elastography (SE) that measures the tissue's stiffness and ability to deform directly by using a mechanical pulse and shear wave elastography (SWE) that measures the ability of a sound wave to propagate within a tissue based on the tissue's elasticity.

4.6 MRI Prostate

MRI is the investigation of choice for assessment of prostate gland and its surrounding structures. With advances in technology multi parametric MRI (mpMRI) was developed which combines the conventional T2 imaging with functional and physiological assessment like diffusion-weighted imaging (DWI) along with apparent diffusion coefficient (ADC) maps and dynamic contrast enhanced (DCE). Previously techniques such as in-vivo MR proton spectroscopy were also included, these are not recommended now due to longer acquisition time and inconsistent results. These techniques have led to a substantial improvement in detection of clinically significant prostate cancer which further aids in reducing mortality and also increasing diagnosis of benign and dormant malignancies in order to reduce unnecessary biopsies and treatment.

PIRADS (Prostate Imaging-Reporting and Data System) v2 was developed by several working groups along with support from ACR and was designed to promote global standardization and reduce the variation in the acquisition, interpretation and reporting of prostate MRI. It was later updated to PIRADS v2.1 to improve detection, localization, characterization and stratification of cases with suspected cancer [3].

Assessment of PIRADS is dependent only on the findings of multiparametric MRI and does not incorporate factors like PSA, DRE or clinical history. It comprises of a 5-point scale that denotes the likelihood or the probability of presence of clinically significant cancer in the visualized prostatic lesions based on the findings on T2 weighted images (T2WI), Diffusion weighted images (DWI) and Dynamic contrast enhanced images (DCE).

PIRADS-1: Very low (clinically significant cancer is highly unlikely)
PIRADS-2: Low (clinically significant cancer unlikely)
PIRADS-3: Intermediate (clinically significant cancer presence is equivocal)
PIRADS-4: High (clinically significant cancer is likely)
PIRADS-5: Very high (clinically significant caner is highly likely)

4.6.1 Timing of MRI

Following TRUS biopsy, changes of hemorrhage and inflammation may be present in the peripheral zone and seminal vesicles with hemorrhage appearing as T1 hyper intense areas. These changes may affect the interpretation of prostate MRI. Although these changes may persist for months, they usually diminish over time and an interval of at least 6 weeks or longer should be considered between the biopsy and MRI. However, if the MRI is performed following a negative TRUS biopsy, the clinically significant cancer if present is usually located at a site that is separate from the biopsy site, therefore the post-biopsy changes may not affect MRI interpretation.

4.6.2 Patient Preparation

To reduce motion artefacts from bowel peristalsis, anti-spasmodic agents may be used. However, these are not used routinely.

Sometime, a rectal enema administration may be considered if presence of stool is causing artifacts and impairing the image quality.

Few of the studies also recommended that patients refrain from ejaculation for 3 days prior to MRI as it maintains good distention of seminal vesicles.

4.7 Multiparametric MRI

Both conventional T1 and T2W sequences are taken. T1W images are obtained primarily to assess presence of hemorrhage within the prostate gland and seminal vesicles and also to delineate the outline of the gland. They are also used to assess nodal and skeletal metastases predominantly in post intravenous administration of gadolinium based contrast agent [4].

T2W images are used to delineate prostate zonal anatomy, intra-glandular abnormalities and for extension of lesion in surrounding structures like seminal vesicles and lymph nodes.

Diffusion weighted imaging (DWI) reflects the component of random water molecules. It consists of high b-values images and ADC map.

The updated version of PIRADS state that the diffusion weighted imaging (DWI) should include at least two sets of b-values, one with low b-value of 0–100 s/mm^2 and one with a high b-value of at least 1400 s/mm^2 as they preserve the signal in areas with restricted diffusion when compared to surrounding normal tissue that show diminished signals due to greater movement between application of gradients with differing b-values.

Most clinically significant cancers which show restricted diffusion appear hypointense on ADC maps. The ADC values above 900 mm^2/s are considered to be likely benign and those below 750 mm^2/s are likely to be malignant. There has been a considerable overlap in ADC values of BPH, low grade and high grade cancers. Also the ADC values are affected by b-values thereby showing inconsistent results, therefore the qualitative visual assessment of ADC map is mostly used as a primary method to assess ADC more than the quantitative values.

Some BPH nodules in transition zone are not clearly encapsulated and may also exhibit hyper intensity on high b-value DWI with corresponding hypo intensity on ADC maps.

4.7.1 Dynamic Contrast Enhanced (DCE) MRI

Rapid acquisitions of T1w gradient echo sequences before, during and after administration of gadolinium based contrast agents. Like other malignant lesions, prostate cancers also show early contrast enhancement when compared with normal tissue. In PIRADS v2.1, “positive” DCE refers to the lesion showing focal early enhancement with corresponding T2W and DWI abnormalities while “negative” DCE is a lesion that either does not enhance early compared to surrounding normal prostate tissue or there is diffuse enhancement so the margins of enhancing area do not correspond on T2 or DWI. However, since there is variability in enhancement characteristics with few showing early washout while others may retain contrast longer, enhancement alone is not considered definitive for diagnosing clinically significant prostate cancer and corresponding T2w and DWI images should be carefully evaluated. Thus, although DCE is an essential component of multi parametric MRI, its role in determining PIRADS v2.1 score is secondary to T2W and DWI sequences (Table 4.1).

Table 4.1 T1 weighted images

Score	Imaging features
Negative	No early or contemporaneous enhancement, or diffuse multifocal enhancement which does not correspond to focal abnormalities on T2W or DWI or focal enhancement corresponding to a lesion showing features of BPH in transition zone on T2
Positive	Focal with early or contemporaneous enhancement compared to normal prostatic tissue corresponding to abnormality on T2W and DWI sequences

Table 4.2 T2 weighted images

DWI imaging appearance	DWI score	DCE	PIRADS category (assessing likelihood of prostate cancer)
No abnormality on ADC map or high b-value	1	NA	1 (Highly unlikely)
Linear or wedge shaped hyperintensity at high b-value DWI **OR** hypointensity on ADC maps	2	NA	2 (Unlikely)
focal discrete hyperintensity on high b-value DWI **OR** hypointensity on ADC map different from background gland **OR** showing markedly hypreintense appearance on DWI **OR** hypointensity on ADC, but not both	3	Negative Positive	3 (Equivocal) 4 (Likely)
Focal markedly hyperintense signal on high b-value DWI and hypointense signal on ADC measuring <1.5 cm in greatest dimension	4	NA	4 (Likely)
Similar as 4 but measures >1.5 cm in greatest dimension or definitive extraprostatic extension or invasive behaviour	5	NA	5 (Highly likely)

4.8 Assessment of Peripheral Zone

The DWI sequence is the predominant sequence to assess the peripheral zone, however few features of T2 imaging findings (like T2 hypo intensity) needs further evaluation with careful interrogation of DWI-ADC images. The complementary sequence to DWI is DCE (Table 4.2).

For peripheral zone lesions, T2W score is used for PIRADS assessment only if the DWI is inadequate or absent (Table 4.3).

4.9 Assessment of Transition Zone

Benign prostatic hypertrophy demonstrates multiple hyperplastic nodules with intervening normal tissue in transition zone. Therefore, it becomes necessary to understand which nodules should be scored. Focal lesions or nodules which showcase features associated with malignancy on T2W OR DWI, thereby differing from background gland should be scored. Therefore, lesions showing greater restricted

Table 4.3 PIRADS score of T2 weighted images

T2W score	T2W imaging characteristic
1	Normal (uniform hyperintense signal intensity)
2	Wedge shaped or linear hypointensity or diffuse mild hypointensity showing indistinct margins
3	Heterogenous signal intensity or rounded non-circumscribed moderately hypointense signal which do not qualify for 2, 4 or 5 scores
4	Homogenous, circumscribed, moderately hypointense nodule <1.5 cm
5	Homogenous, circumscribed, moderately hypointense nodule >1.5 cm or definitive invasive behaviour/extra-prostatic extension

Table 4.4 Chances of cancer prostate—according to T2 score and DWI score

T2W imaging characteristic	T2W score	DWI appearance (DWI score)	PIRADS category (assessing chances of prostate cancer)
Normal appearing transition zone or completely encapsulated round nodule—typical nodule	1	NA	1 (Highly unlikely)
Predominantly encapsulated nodule or unencapsulated nodule with homogenous circumscribed appearance (atypical nodule) or mildly hypointense homogenous area between nodules	2	<4 ≥4	2 (Unlikely) 3 (Equivocal)
Heterogenous signal intensity showing obscured margins and those who do not qualify in other T2 scores	3	<5 5	3 (Equivocal) 4 (Likely)
Non-circumscribed or lenticular shaped lesion showing homogenous moderately hypointense signal intensity that is <1.5 cm in greatest dimension	4	NA	4 (Likely)
Similar to 4 but >1.5 cm in greatest dimension or shows invasive behaviour or features of definitive extraprostatic extension	5	NA	5 (Highly likely)

diffusion than the background gland or having obscured margins or invasive features on T2W images should be scored.

The T2W imaging characteristics is the dominant factor in assessing the PIRADS category of nodules in transition zone. Although T2W is the dominant factor in transition zone, restricted diffusion is also an important characteristic. Therefore, in atypical transition zone nodules with a T2 score of 2 are upgraded to PIRADS category 3 of they have a DWI score of more than 4 (i.e. marked diffusion restriction) and of T2 score of 3 to be upgraded to 4 if they have DWI score of 5.

Mild to moderately restricting nodules may also be seen in encapsulated lesions in transition zone which may represent stromal hyperplasia should not be upgraded, therefore nodule with T2 score of 1 should not be upgraded to PIRADS 2 based on DWI category 3 (mildly/moderately restricting) (Table 4.4).

4.9.1 Central Zone

The normal central zone is usually visible as bilaterally symmetrical low signal intensity tissue on both T2W and ADC images which surround the ejaculatory ducts extending from prostatic base to verumontanum. It shows mild hyper intense signal on high b-value DWI with no evidence of early enhancement. Incidence of prostate cancers in this zone is uncommon and usually extends from adjacent peripheral or transitional zones. Focal asymmetry with T2/DWI abnormalities or early enhancement may be suspicious for presence of malignancy. However, it should be noted that BPH may also deform, displace or cause focal asymmetry of the central zone.

4.9.2 Anterior Fibromuscular Stroma

Seen as crescentic symmetrical low signal intensity region on T2 (similar to pelvic floor muscles), ADC and high b-value DWI showing no obvious early enhancement. Hyperintense signal intensity on T2/DWI with corresponding low signal on ADC, focal enlargement with evidence of early contrast enhancement may help to detect malignant lesions that may have extended to anterior fibromuscular stroma (Fig. 4.5).

4.9.3 Extraprostatic Extension

Prostate lacks a true capsule, however there is presence of thin, dark rim on T2 images that partially surround the prostate gland forming the pseudo-capsule which usually comprises of bands of concentric fibro-muscular tissue inseparable from prostatic stroma. It is usually incomplete anteriorly and apically. Features which suggest presence of extra-prostatic extension include, (Fig. 4.2a, b)

- Frank capsular breech by the suspicious lesion with possible invasion of adjacent bladder or rectum.
- Broad capsular contact (10–20 mm)
- Irregular, spiculated or angulated prostatic margin adjacent to tumor or presence of focal bulge of capsule.
- Asymmetry, thickening or traction of adjacent neuromuscular bundles.
- Invasion of seminal vesicles which usually present as loss of normal hyperintense signal intensity on T2 or abnormal contrast enhancement/diffusion restriction. Also involvement can be suspected when the angle between the prostate base and seminal vesicle is obliterated (Fig. 4.6).

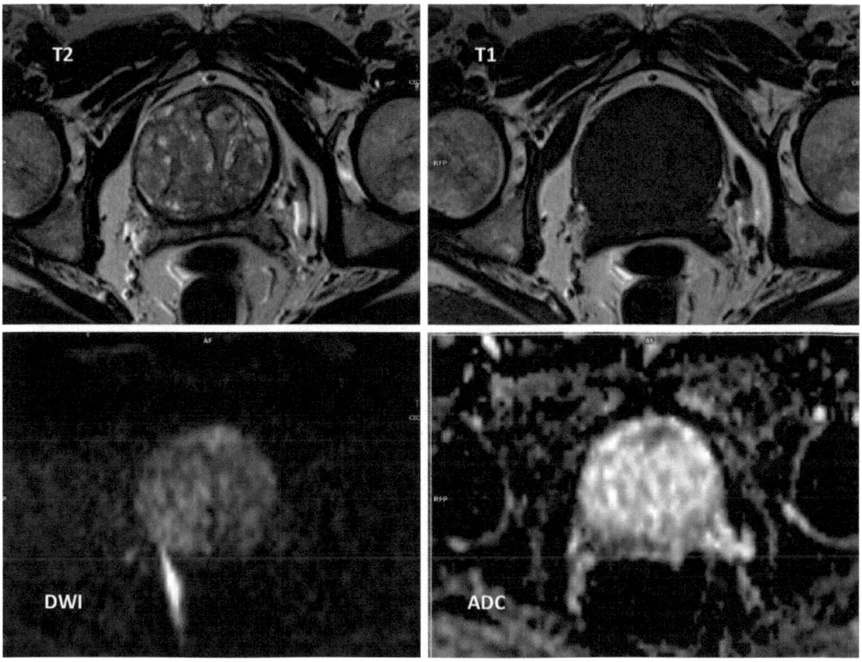

Fig. 4.5 Anterior fibromuscular stroma. Small T2 hypointense mass lesion in left mid anterior fibro muscular region showing restricted diffusion (source—self)

4.9.4 Mapping Lesions on MRI

Since prostate cancers are usually multifocal, the largest tumor with the highest Gleason score usually is the predominant contributing factor in extra-prostatic extension and positive surgical margins.

In cases with up to four lesions, the index lesion is identified as the one which has the highest PIRADS category. If the highest category is shared among two lesions, the index lesion is the one that shows extra-prostatic extension. Thus, a lesion showing extra-prostatic extension will be identified as the index lesion despite the presence of a larger tumor if both of them have the same PIRADS category.

If there are more than four suspicious lesions then only four which have highest likelihood of clinically significant malignancy (highest PIRADS category) should be included in the report.

Largest dimension of a lesion is usually taken in axial images, however if the largest dimension is in sagittal or coronal images, that should also be reported. In peripheral zone, the lesions should be measured in ADC and in transition zone, they should be measured on T2.

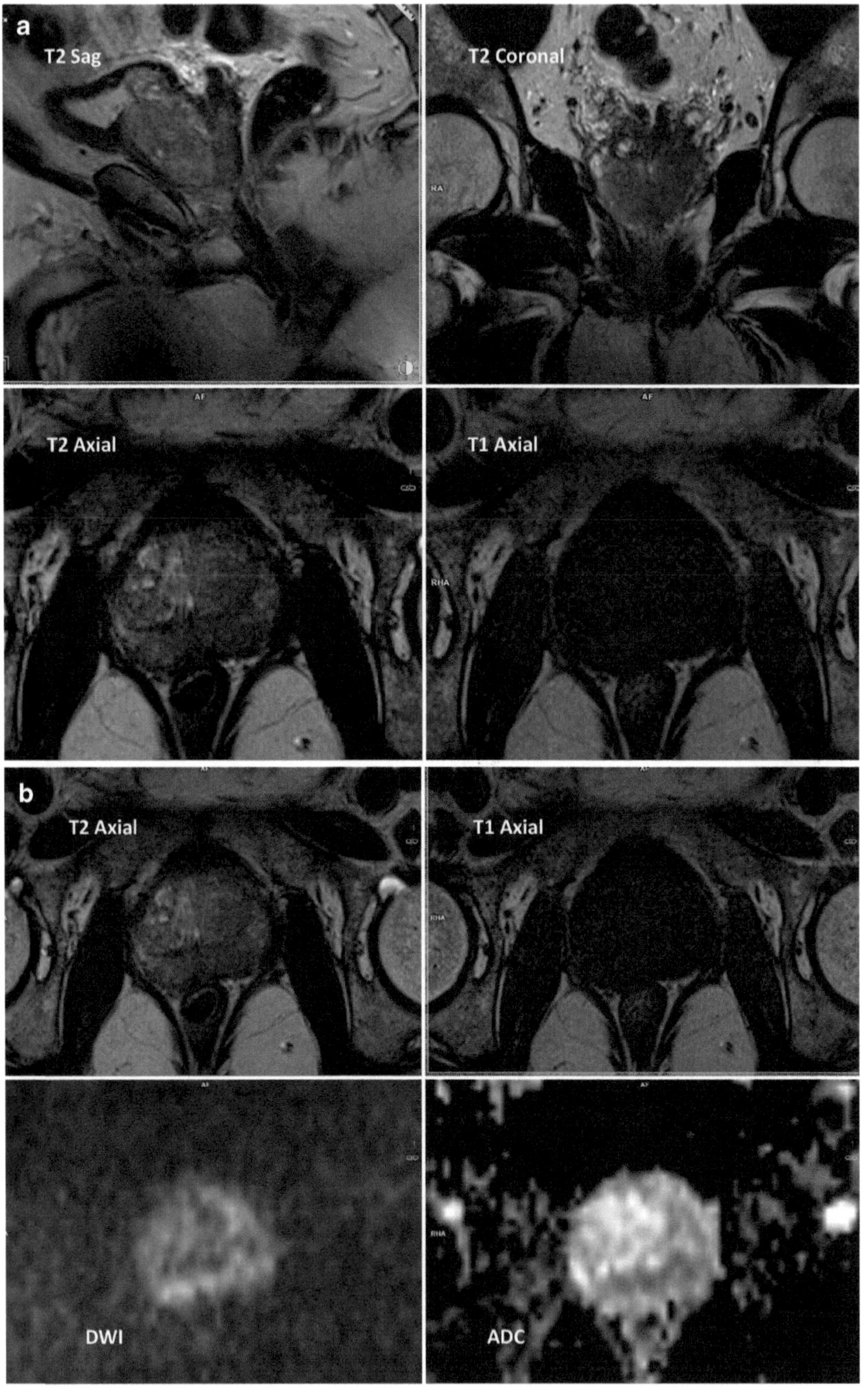

Fig.4.6 (**a**) Large T2and T1 hypointense mass lesion involving the posterior peripheral zone, extending from base to the apex with extracapsular extension posteriorly PIRADS 5 (source—self). (**b**) Large T2 and T1 hypointense mass lesion involving the posterior peripheral zone, extending from base to the apex with extracapsular extension PIRADS 5 (source—self)

4.9.5 Sector Map

Segmentation model of PIRADS v2.1 evaluates 41 sectors/regions.

38 sectors are from prostate gland wherein the gland is divided into right and left by a vertical line drawn along the prostatic urethra and in anterior and posterior parts using horizontal line through the middle of gland.

The right and left peripheral zones are divided into three sections, namely anterior (a), medial posterior (mp) and lateral posterior (lp) at base, mid gland and apex regions.

The right and left transitional zones are divided into two sections, namely anterior (a), and posterior (p) at base, mid gland and apex regions.

The anterior fibromuscular stroma is divided into right and left zones at base, mid gland and apex regions.

The seminal vesicles constitute two sectors and are divided into right and left.

One sector is for the external urethral sphincter.

Benign prostatic lesions:

Many of the prostatic abnormalities in MRI can be benign. A few of them are detailed below:

4.10 Benign Prostatic Hyperplasia

With advancing age, there is occurrence of benign prostatic hyperplasia that usually develops in response to testosterone after its conversion to dihydrotestosterone. Most often, the BPH nodules arise from transition zone with few of them extruding into peripheral or central zone. It consists of a mixture of stroma and glandular hyper-plastic tissue appearing on MRI as band-like areas and/or encapsulated nodules which along with cystic atrophy of the gland typically demonstrate moderate to marked T2 hyper intensity unlike the malignant lesions and may show high vascularity on DCE with variable signals on DWI.

- **Prostatitis**

- It is the inflammation of prostate gland which can show decreased signal in peripheral zone on both T2 and ADC which appear as band or wedge shaped or diffuse. Since it also increases perfusion it can show a "false-positive" DCE result.
- **Atrophy**
- Commonly seen a sequelae to old age or chronic inflammation and is seen as wedge shaped areas of low T2 signal intensity with subtle to mild decrease in ADC map with associated contour retraction of the gland. Fibrosis can also occur post inflammation and is seen as band or wedge shaped areas of low T2 signal intensity.
- **Hemorrhage**

- It is commonly seen after biopsy mostly in peripheral zones and/or seminal vesicles. It shows diffuse hyper intense signal on T1W images and iso-hypointense signal on T2. Chronic blood products appear hypo intense on all sequences.
- **Cysts**
- Show hyper intensity on T2 and hypo intensity on T1 denoting their fluid content. They can also contain hemorrhage or proteinaceous content which is demonstrated as T1 hyper intense signal.
- **Calcifications**
- Appear markedly hypo intense on all pulse echo sequences.

4.11 PET-CT for Staging of Prostate Cancer

Prostate-specific membrane antigen (PSMA) targeted radiopharmaceuticals play a large role for initial staging and evaluation of biochemical recurrence for locally advanced prostate cancer. With a higher detection rate and superior diagnostic ability when compared with conventional imaging, PSMA-PET can also impact the patient treatment. Gallium 68 (Ga^{68} PSma-11) PET uptake correlates with PSMA level expression in prostate cancer cells in-vivo.

4.12 Lymph Node Metastases

Multiparametric MRI and PET-CT are the primary diagnostic tools for assessing lymph node metastasis. In MRI, DWI is the best sequence to evaluate the lymph nodes. The suspicious characteristics of lymph nodes which may suggest metastasis include round shape with short axis diameter $\geq$8 mm or oval shape with short axis diameter $\geq$10 mm, heterogenous shape with irregular margins. However, conventional MRI cannot detect micro-metastasis. Therefore, MR lymphangiography using ultra-small super paramagnetic iron oxide (USPIO) has shown high specificity and sensitivity in detection of normal sized lymph nodes with micro-metastasis. These USPIO are normally taken up by the macrophages in normal lymph node creating a low signal in normal areas. Nodal areas devoid of macrophages show no loss of signal, thereby signifying presence of disease.

Lymph nodes can be characterized as regional (N1 stage) namely, pelvic, hypogastric, sacral and iliac (internal and external) levels.

Distant lymph nodes which include common iliac, aortic, inguinal and supraclavicular are staged as metastatic disease stage M1a.

4.12.1 Skeletal Metastasis

These occur in majority of patients with clinically significant prostate cancer and the burden of the bone disease directly corresponds to survival. This spread follows the 'seed-and-soil' hypothesis where lesions are mostly located in the vertebra and ribs due to dissemination of the disease via the baton's venous plexus. The spread of disease in the bone follows the distribution of adult red bone marrow i.e. skull, thorax, pelvis, spine, proximal long bones and eventually spreads to adjacent cortical bone. Although predominantly osteoblastic, few of the metastatic lesions can also be osteolytic or mixed.

Imaging for evaluation of skeletal metastasis may involve a number of imaging modalities with Tc^{99} MDP (methylene bisphosphonate) bone scintigraphy forming the backbone of detection of skeletal metastasis. It is backed by PET-CT and MRI.

References

1. Sklinda K, Frączek M, Mruk B, Walecki J. Normal 3T MR anatomy of the prostate gland and surrounding structures. Adv Med. 2019;2019:3040859. https://doi.org/10.1155/2019/3040859. PMID: 31276002; PMCID: PMC6558623.
2. Tyloch JF, Wieczorek AP. The standards of an ultrasound examination of the prostate gland. Part 1. J Ultrason. 2016;16(67):378–90. https://doi.org/10.15557/JoU.2016.0038. Epub 2016 Dec 30. PMID: 28138409; PMCID: PMC5269525.
3. Purysko AS, Rosenkrantz AB, Turkbey IB, Macura KJ. Radiographics update: PI-RADS version 2.1—a pictorial update. Radiographics. 2020;40(7):E33–7.
4. Ludwig DR, Fraum TJ, Fowler KJ, Ippolito JE. Imaging in prostate cancer: magnetic resonance imaging and beyond. Mol Med. 2018;115(2):135–41. PMID: 30228705; PMCID: PMC6139856.

Chapter 5
Understanding Gleason Grading of Prostate Cancer

Swapnil Rane and Santosh Menon

Gleason grading suggest aggressiveness of the Cancer. Gleason grading is important and according to Gleason grading further treatment modality and prognosis of the disease helps in management and counselling of the patient.

5.1 Basic Introduction to Prostate Cancer Pathology

Prostatic adenocarcinoma, in simple terms, refers to prostate cancer. Adenocarcinoma specifically refers to a cancer that arises from the glandular cells of an organ. In the case of prostatic adenocarcinoma, the cancerous cells originate in the glandular tissue of the prostate.

Usually this cancer grows slowly, and early-stage prostate cancer may not cause noticeable symptoms. However, as the cancer progresses, it can lead to urinary problems such as frequent urination, weak urine flow etc.

The confirmatory diagnosis of prostate cancer is on a biopsy, commonly called as a needle core biopsy. This biopsy is stained and taken onto a glass slide and seen under a microscope by a pathologist, who diagnose the cancer. Under the microscope the pathologist looks at certain features of the malignant glands like smaller size compared to normal glands, rigid shape of glands, and certain nuclear features like prominence of nucleoli in the cells of malignant glands. As discussed below, further, the pathologist will grade/score these cancers based on the pattern of the malignant glands.

S. Rane · S. Menon (✉)
Department of Pathology, Tata Memorial Centre, Mumbai, India

N. Gupta (ed.), *A Clinical Guide for Cancer Prostate Patient*,
https://doi.org/10.1007/978-981-97-5357-4_5

When diagnosed with prostate cancer, doctors use various tools to determine the severity and aggressiveness of the disease. One important method used is called Gleason grading. In this chapter, we will explain what Gleason grading is, how it works, and what it means for individuals diagnosed with prostate cancer [1, 2].

5.2 Gleason Grading [3–6]

- Gleason grading is a system used by pathologists to assess the microscopic appearance of prostate cancer cells under a microscope. It helps determine the aggressiveness of the cancer by evaluating the pattern and architecture of the tumour cells. The grading system is named after Dr. Donald Gleason, who developed it in the 1960s.

5.2.1 The Gleason Score

- This is done either on a needle core biopsy from the prostate before surgery and is also done after prostate gland has been operated for a cancer. The Gleason grading system assigns a score to prostate cancer based on its microscopic appearance. The system has undergone changes since its inception.
- In is original form, the appearance/tumor arrangement and pattern are assigned a score starting from 1 to 5 depending on how much it resembles normal prostate gland (1 being very similar to normal/least aggressive and 5 being least similar/most aggressive).
- The primary and secondary patterns represent the two most common types of cancer cell patterns observed in the biopsy sample. The primary pattern is the predominant type, while the secondary pattern is the second most common type.
- The Gleason score is calculated by adding the primary (the most common) and secondary (the second most common) patterns observed in the tumour. For example, if the primary pattern is graded as 3 and the secondary pattern is graded as 4, the Gleason score would be 7 (3 + 4). The Gleason score ranges from 2 to 10, with 2 being the least aggressive and 10 being the most aggressive form of cancer.

5.2.2 Interpreting the Gleason Score

- The Gleason score provides valuable information about the aggressiveness of the cancer and helps guide treatment decisions. Lower scores (6 or less) indicate less aggressive cancer, which may require less aggressive treatment or active surveillance. Scores between 7 and 10 suggest more aggressive cancer, which may necessitate more aggressive treatment options, such as surgery, radiation, or hormone therapy.

5.2.3 Clinical Implications

- Gleason grading helps doctors predict the likelihood of cancer spreading beyond the prostate gland and provides insights into the prognosis. Higher Gleason scores indicate a higher risk of cancer spreading to other parts of the body. Therefore, individuals with higher scores may require more intensive treatment and close monitoring to prevent the cancer from spreading beyond the prostate. Unfortunately, as is with other cancers, many prostate cancer patients present when it has already spread to bones, lymph glands, liver and other organs. In general in such scenarios, the Gleason score is higher and almost always more than score 6.

5.2.4 Limitations of Gleason Grading

- While Gleason grading is a valuable tool, it is important to note its limitations. The grading system relies on subjective interpretation by pathologists, which can lead to some variability in results. Additionally, the grading is based on a small sample of the tumor obtained through biopsy, and the actual extent and characteristics of the cancer may vary within the prostate gland.

5.2.5 Advances in Gleason Grading

- Over the years, medical researchers and practitioners have continuously strived to improve the accuracy and effectiveness of diagnostic tools and grading systems. One such advancement is the modified Gleason grading system, an enhanced approach to assessing prostate cancer aggressiveness.

5.2.6 The Need for Modified Gleason Grading

- The modified Gleason grading system was introduced in 2005 after consensus by the International Society for Urologic Pathology (ISUP) to overcome the shortcomings of the original system. It recognized the need to incorporate additional information, such as the percentage of cancer involvement within each pattern and the presence of tertiary patterns, to improve the accuracy of the grading process. This modification aimed to provide a more comprehensive evaluation of the tumour and improve the predictability of how the cancer may progress.

5.2.7 Key Features of Modified Gleason Grading

- Incorporation of highest grade: It was slowly, over the years, recognized that the presence of very scanty percentage of a highest grade pattern in a needle core biopsy, influenced a worser outcome of patient. Hence, the modified Gleason score now includes the most predominant added to the highest grade pattern.
- Quantification of tumour involvement: The modified Gleason grading system emphasizes quantifying the percentage of cancer involvement within each pattern. This information provides a clearer understanding of the tumour's composition, allowing for a more accurate determination of aggressiveness.

5.2.8 Refinements to the Grading Scale

- Over time, it was realised that Patterns 1 and 2 as defined in the original Gleason score were found to be very rare and commonly represented mimics of cancer rather than true cancer as per modern understandings. Hence, these patterns were eliminated from the Gleason system. The Modified Gleason score now ranged from 6 to 10 instead of 2 to 10.
- It also introduced separation of Gleason score 7 into two further categories ($3 + 4 = 7, 4 + 3 = 7$) to better differentiate between tumours with varying degrees of aggressiveness, offering clinicians and patients more precise prognostic information.

5.2.9 Advantages of Modified Gleason Grading

- Improved accuracy: By incorporating highest grade patterns and quantifying tumour involvement, the modified Gleason grading system provides a more nuanced and accurate assessment of prostate cancer aggressiveness. This refined evaluation enables clinicians to make more informed treatment decisions tailored to each patient's specific needs.

- Enhanced prognostic value: The modified grading system offers enhanced prognostic capabilities, providing better predictions of disease progression and outcomes. This allows the treating urologist/uro-oncologist to stratify patients into appropriate risk groups, guiding the selection of treatment modalities and intensity. It enables the identification of patients at higher risk of disease progression and metastasis. This information allows for the selection of appropriate treatment strategies, including active surveillance, surgery, radiation therapy, or hormonal therapies, based on the individual patient's risk profile. The refined grading system supports personalized medicine approaches, optimizing patient outcomes.
- Standardization and reproducibility: The modified Gleason grading system promotes greater standardization and reproducibility across different institutions and pathologists. The inclusion of detailed guidelines and criteria assists pathologists in making consistent assessments, minimizing variability in the interpretation of biopsy samples.
- Research and Validation: The modified Gleason grading system has been extensively researched and validated through studies and collaborations among pathologists, urologists, and oncologists. These efforts have aimed to refine the grading criteria, establish consensus guidelines, and ensure the system's accuracy and reproducibility. Ongoing research continues to refine and validate the modified grading system, aiming to improve its applicability and effectiveness in clinical practice.
- It is important to note that the adoption of the modified Gleason grading system may vary across different institutions and regions. However, the overall goal remains consistent—to provide a more accurate assessment of prostate cancer aggressiveness, enhance prognostic capabilities, and guide personalized treatment decisions for improved patient outcomes.

5.2.10 Gleason Grade Grouping

- In 2014, International Society of Urological Pathology Consensus Conference on Gleason Grading of Prostatic Carcinoma suggested grade groupings. Then this was validated in a multi-institutional study, includes five distinct Grade Groups based on the modified Gleason score groups.
- Grade Group 1 = Gleason score ≤6,
- Grade Group 2 = Gleason score 3 + 4 = 7,
- Grade Group 3 = Gleason score 4 + 3 = 7,
- Grade Group 4 = Gleason score 8,
- Grade Group 5 = Gleason scores 9 and 10.
- Currently, grade grouping is used along with Gleason scoring for the treatment as well as for the counselling of the patient.

5.3 Conclusion [5, 6]

- Gleason grading is a crucial aspect of diagnosing and understanding prostate cancer. It helps doctors determine the aggressiveness of the cancer, guide treatment decisions, and predict the prognosis. Through improved standardization and reproducibility, clinicians can make better-informed treatment decisions, offering patients tailored approaches that maximize therapeutic benefits. As research continues to evolve, modified Gleason grading stands as a testament to the ongoing efforts to enhance diagnostic accuracy and improve patient outcomes in prostate cancer care.
- By considering the Gleason score, medical professionals can provide personalized care and develop treatment plans tailored to the individual's need.

References

1. Baraban E, Epstein J. Prostate cancer: update on grading and reporting. Surg Pathol Clin. 2022;15(4):579–89. https://doi.org/10.1016/j.path.2022.07.008. Epub 2022 Oct 13. PMID: 36344176 Review.
2. Humphrey PA. Histopathology of prostate cancer. Cold Spring Harb Perspect Med. 2017;7(10):a030411. https://doi.org/10.1101/cshperspect.a030411. PMID: 28389514 free PMC article. Review.
3. Kweldam CF, van Leenders GJ, van der Kwast T. Grading of prostate cancer: a work in progress. Histopathology. 2019;74(1):146–60. https://doi.org/10.1111/his.13767. PMID: 30565302 free PMC article. Review.
4. Kryvenko ON, Epstein JI. Changes in prostate cancer grading: including a new patient-centric grading system. Prostate. 2016;76(5):427–33. https://doi.org/10.1002/pros.23142. Epub 2015 Dec 28. PMID: 26709152 Review.
5. Epstein JI, Egevad L, Amin MB, Delahunt B, Srigley JR, Humphrey PA, Grading Committee. The 2014 International Society of Urological Pathology (ISUP) Consensus Conference on Gleason Grading of Prostatic Carcinoma: Definition of grading patterns and proposal for a new grading system. Am J Surg Pathol. 2016;40(2):244–52. https://doi.org/10.1097/PAS.0000000000000530. PMID: 26492179 Review.
6. Kryvenko ON, Epstein JI. Prostate cancer grading: a decade after the 2005 modified Gleason grading system. Arch Pathol Lab Med. 2016;140(10):1140–52. https://doi.org/10.5858/arpa.2015-0487-SA. Epub 2016 Jan 12. PMID: 26756649.

Chapter 6
Prostate Cancer Risk Stratification

Rishi Nayyar and Siddharth Jain

"Life is 10% what happens to you and 90% how you react to it"
Charles. R. Swindoll
"Life isn't about waiting for storm to pass, it's about learning how to dance in rain"

Vivian Greene

6.1 Introduction

- Getting diagnosed with a cancer is a life changing event. It affects not only medically and physically but also emotionally and financially. Its effect is not only perceived by the patient but also by the family members. However, in the current age with availability of robotics, genetics, cutting edge nuclear medicine, radiation machines and artificial intelligence many treatment options are available which can be tailored according to needs of each individual patient. Therefore, by understanding one's condition and participating in decision making one can best direct the treatment of disease for oneself as per wishes.

6.2 What Is Cancer Risk Stratification?

- Once the diagnosis of cancer is confirmed on biopsy test, next step is to decide treatment for it. This is based broadly on two parameters. First, what is the present extent to which cancer has already spread. This is called as 'Staging'. Second, what is the likelihood of cancer to progress and cause harm. This is called as 'Risk stratification'.

R. Nayyar (✉) · S. Jain
Department of Urology, All India Institute of Medical Sciences, New Delhi, New Delhi, India

N. Gupta (ed.), *A Clinical Guide for Cancer Prostate Patient*,
https://doi.org/10.1007/978-981-97-5357-4_6

- In cancer management various physical, imaging based, biochemical, or nomogram scorebased criteria are used for risk stratification to decide best course of treatment among available treatment options based on likelihood of progression of disease or harm from it. These criteria may be variable for different cancers and are selected based on available data from various medical scientific studies. By understanding risk stratification, the outcome of various treatments in form of cancer eradication, control, recurrence, progression and cancer related mortality can be predicted to a reasonable extent.

6.3 What All Parameters are Included in Risk Stratification of Prostate Cancer?

- Prominent cancer guideline forming medical societies (like the Union for International Cancer Control, American Joint Committee on Cancer, European association of urologist and others) use following parameters for risk stratification of Prostate cancer.
- Clinical tumor staging: While staging is an important parameter to decide treatment by itself, it is also an important component for risk stratification of prostate cancer. Staging includes local tumor spread (extent to which prostate gland and structures around it are involved) called as T staging, lymph node spread (extent to which lymphatic tissue in the body is involved) called as N staging and staging of metastatic (involvement of rest of bodily organs) deposits in body called M staging (Fig. 6.1).

6.4 T Staging Is Further Divided into T1 to T4

- T1: Clinically non palpable prostate cancer on rectal examination.
- T2: Tumor involves prostate gland only. It is further sub divided into T2a when one half of one side lobe is involved, T2b when more than one half of one lobe is involved. T2c when tumor involves both lobes of prostate.
- T3: Tumor extends beyond the prostate gland but is not fixed or does not invade adjacent structures.
- T4: Tumor is fixed or invades adjacent structures other than seminal vesicles, such as sphincter muscles, rectum, bladder, levator muscles, and/or pelvic wall.

6.4.1 N Staging

- N1: metastasis to regional pelvic lymph nodes only.

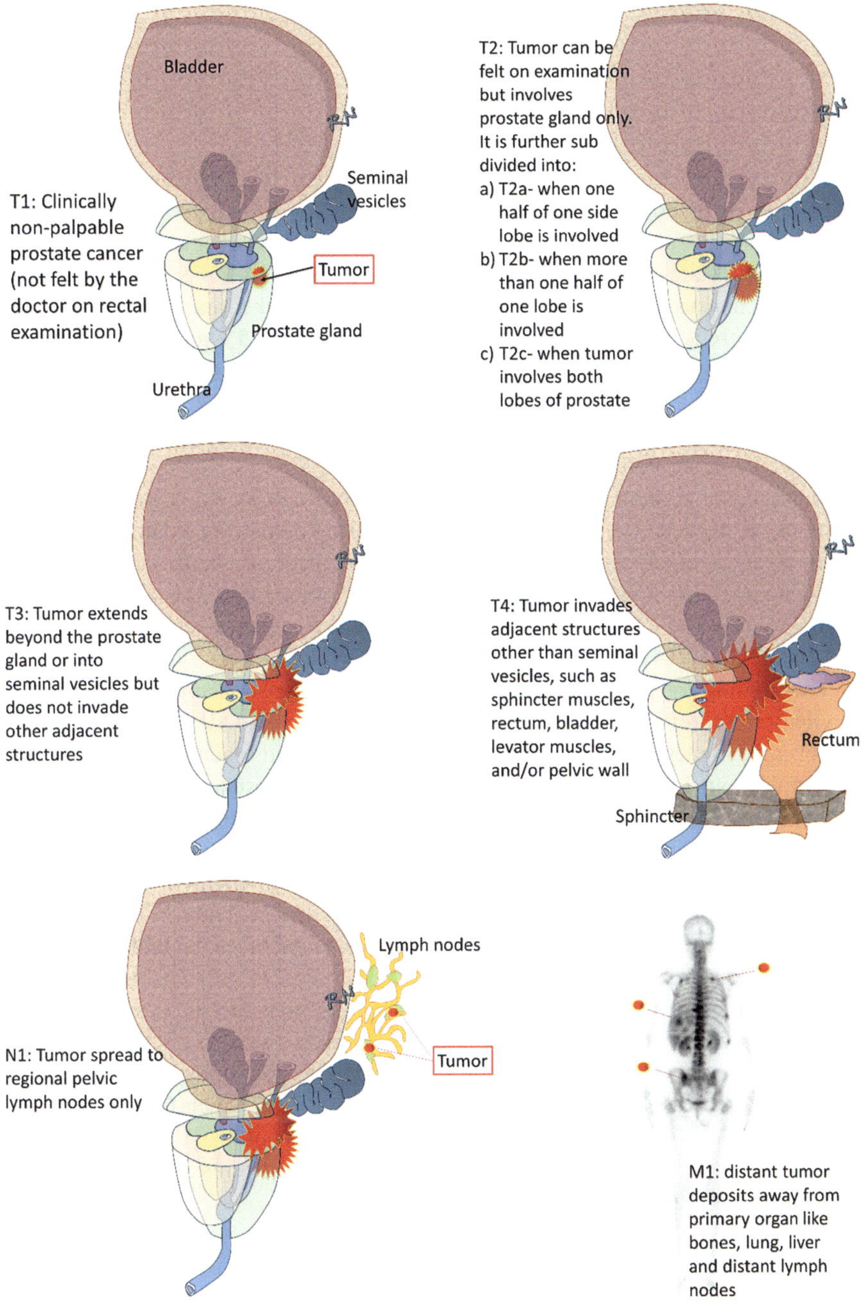

Fig. 6.1 Illustrative description of staging of prostate cancer (TNM staging)

6.4.2 M Staging

- M1: distant tumor deposits away from primary organ like bones, lung, liver and distant lymph nodes [1–3].

- PSA levels: PSA stands for prostate specific antigen. It is a biochemical secreted by prostate gland in semen and blood. If there is any pathology ongoing in prostate gland, PSA levels in blood increase. Thus, PSA is marker of prostate pathology. In prostate cancer, blood PSA levels are increased. Higher the level, higher are the chances of having prostate cancer and higher risk disease also. In absolute terms, there is a risk of having metastases in about 2%, 5% and 16% of patients with PSA levels less than 10, 10–20 and 20–50 ng/mL respectively, among men diagnosed of having prostate cancer. Correspondingly, one could conclude that the risks of failure of local treatment and presence or development of new metastasis are higher with increasing values of PSA [4, 5].
- Pathological grading on prostate biopsy: To confirm existence of prostate cancer biopsy is needed. The tissue samples taken during biopsy are examined not only for presence of cancer but also for the grade of cancerous changes. This grade of a cancer tells us how abnormal does the cancer tissue looks as compared to normal tissues, and can give the doctor an idea of how aggressive the cancer is and what treatment is needed. (Details in other chapter).
- The grading for prostate cancer is done by pathologists using a system called Gleason score system, named after a pathologist who devised this system. Several samples of prostate biopsies are examined and grades are given to each sample from 1 to 5. Grades 1 and 2 are near normal prostate gland cells. Grades 3–5 are definitive prostate cancer cells. Grade 5 being worst and most rapidly multiplying and spreading type of cells. The two highest grades found on a biopsy are summed up together to calculate the overall Gleason score, which if lies in range of 6–10 means presence of prostate cancer. Thus, in this system, a prostate cancer with score of 6 is least dangerous type of cancer and that with 10 is most dangerous type.
- For avoiding confusion with scores of 2–5 being misunderstood as cancer and better grouping of cases having similar type of risk, a system of Gleason grouping has been developed. Grade group 1 is when Gleason score is 6, Grade group 2 is when Gleason score is 7 (3 + 4). Grade group 3 is when Gleason score is unfavorable 7 (4 + 3). Grade group 4 is Gleason score of 8 and Grade group 5 is Gleason score of 9 or 10. Grade group 4–5 are real aggressive type requiring aggressive management [6, 7].

6.5 Why Is Risk Stratification Needed?

- All prostate cancers do not behave in a same manner. A lower grade, lower stage and low PSA cancer can be totally insignificant and will not trouble at all in any manner over life. On the other hand, a higher grade, higher stage and high PSA cancer may be life threatening as it tends to grow and spread very fast. Even cancers presenting at same stage as per staging system do not behave same in all cases. At the same time, in this age there are many different type of management options that are available to treat prostate cancer like active surveillance (doing no treatment but only monitoring), surgical removal of prostate gland, radiotherapy, hormone therapy, chemotherapy, focal ablation therapy, bone therapy, radionuclide therapy, immunotherapy and any combination of these. With so many available treatment options and different types of prostate cancer behaviors, the only way forward for deciding treatment for a particular patient is to match the suitable treatment strategy to the specific risk category of that patient so that best possible cancer control is achieved with least exposure to side effects of therapy.

6.6 What are Prostate Cancer Risk Stratification Categories?

- Based on the various tests done including serum PSA (blood test), MRI, CT scan, Bone scan, PET scan and biopsy findings prostate cancer behavior is currently categorized into the following groups:
- Very-low-risk group: These prostate cancers are small, not felt on exam, can only be found in a small area of the prostate, and have not grown outside the prostate which is T stage 1. These patients will have a Grade Group of 1 which is Gleason score of 6 and low blood PSA levels (less than 10 ng/mL) and few other classification criteria. They usually grow very slowly and are unlikely to ever cause any symptoms or other health problems.
- Low-risk group: Prostate cancers in this group of patients will not have yet grown outside of the prostate, have a Grade Group of 1 (Gleason score of 6) and low PSA levels (less than 10), but have slightly larger tumor burden which can be felt on rectal examination involving less than half of one side lobe of prostate gland T2a.
- Intermediate-risk group: This group of prostate cancers can be felt on exam or can be seen on an imaging test. The cancer might be found in more than half of one side of the prostate (T2b) or in both sides of the prostate (T2c), or have a Grade Group of 2 or 3 (Gleason score of 7) or a PSA level between 10 and 20 ng/mL. Additional classifications split the intermediate-risk group into favorable and unfavorable categories.

- High-risk group Prostate cancers: In this group cancer has grown outside the prostate (T3) or have a Grade Group of 4 or 5 (Gleason score 8–10) or have a PSA level of more than 20.
- Very-high-risk group: These prostate cancers have a very high risk for the tumor growing, coming back, or spreading to other organs. They have one or more of the following traits: The tumor has spread to the seminal vesicles (T3) or into other tissues next to the prostate (T4). The biopsy tissue shows areas with a Gleason 5 pattern (Gleason score 9 or 10 or Grade Group 5). They have 2 or 3 of the features found in the high-risk group. Even if final biopsy after radical prostatectomy (surgical removal of prostate) does not show spread beyond the prostate, this group of patients can have recurrence of cancer at the site of surgery or elsewhere in other parts of the body [1, 2].

6.7 How Should Treatment Be Chosen for Prostate Cancer Based on Risk Stratification?

- All the risk stratification-based groupings of prostate cancer as well as their suggested managements are not disjointed from each other but actually reflect only a continuous spectrum of disease. For each of these groups, there are defined guideline directed management strategies that are followed by the urologists. Very low risk and low-risk prostate cancers can be safely followed with active surveillance strategy. One may choose to adopt surgical treatment or radiotherapy if and when the cancer progresses or at the onset itself after fully understanding the side effects of therapy. Intermediate-risk and high-risk cases should be more receptive for surgical treatment or radiotherapy. Men who are 75 years of age or less and are likely to live for 10 years or more (based on their general health and other co-morbid illnesses like heart disease, blood pressure and diabetes) should adopt surgical treatment as it provides best available modality for cure over the long term. High-risk and very high-risk cases are more likely to end up receiving multiple simultaneous therapies (multimodality), because of their very high risk of progression based on their stratification.

6.8 What are Newer Ways for Risk Stratification?

- It is seen that even patients within high and very high-risk group or some in lower-risk group behave inconsistently for that grouping. With research in field of cancer genetics newer parameters are being developed for better understanding. Cancer cells are body cells that have genetic mutations at level of DNA and now behave abnormally. Through genetic tests many of these mutations have been identified. Currently efforts are being made to see how these mutations affect prostate cancer behavior in large population. It is still early stage for these mutation-based risk stratification but it may hold several answers in future. Some of these mutation tests are now available commercially in some countries but their significance is still to be understood completely. Your doctor may suggest some of these tests if you fall in category of patients who may potentially benefit with these. These tests look at which genes or proteins are active inside the prostate cancer cells taken through biopsy. These are used to create a risk score on a scale. A higher score indicating a cancer is more likely to spread to other parts of the body, which might be helpful in determining the best treatment options. Most of these tests involve laboratory intensive processing of tissue or blood samples.

- Examples of such tests include: Decipher, Oncotype DX Prostate, Prolaris, ProMark.
- Usually, samples are sent to labs offering these tests and hold patents for, therefore these tests involve significant cost.

References

1. Klein EA. Prostate cancer: risk stratification and choice of initial treatment. In: Savarese DMF, editor. UpToDate. Waltham: UpToDate; 2021.
2. National Comprehensive Cancer Network (NCCN). Practice Guidelines in Oncology: Prostate Cancer. Version 1.2022.
3. American Joint Committee on Cancer. Prostate. In: AJCC cancer staging manual. 8th ed. New York: Springer; 2017. p. 715–25.
4. Abuzallouf S, Dayes I, Lukka H. Baseline staging of newly diagnosed prostate cancer: a summary of the literature. J Urol. 2004;171(6 Pt 1):2122–7.
5. Ross A, D'Amico AV, Freedland S. Molecular prognostic tests for prostate cancer. In: Savarese DMF, editor. UpToDate. Waltham: UpToDate; 2021.
6. Epstein JI. An update of the Gleason grading system. J Urol. 2010;183:433–40.
7. Epstein JI, Zelefsky MJ, Sjoberg DD, et al. A contemporary prostate cancer grading system: a validated alternative to the Gleason score. Eur Urol. 2016;69(3):428–35.

Chapter 7
Management of Low-Risk Cancer Prostate

Sumit Agrawal, Siddharth Yadav, and Anup Kumar

Every low risk prostate cancer do not need invasive treatment.

7.1 What Is Low Risk Prostate Cancer?

Low risk prostate cancer is subset of prostate cancer which poses little threat to the patient's life as it is believed to be less aggressive and have a low risk of progression.

Low risk prostate cancer is localized within the prostate and is characterized by [1]:

- Prostate Specific Antigen (PSA) level less than or equal to 10 ng/mL
- And Gleason Score <7 (i.e., 6 or less) (ISUP Grade I)
- And Clinical Stage T1–T2a

7.2 How Does Low Risk Prostate Cancer Affect Life?

Men with low-risk prostate cancer tend to have good life span and a low risk of dying due to the prostate cancer itself. The 10-year prostate cancer specific mortality of men 66–74 years is 4.8%. It means that chance of dying due to prostate cancer in the next 10 years after diagnosis is around 4.8 per 100 people affected [2]. In a recent study, risk of death at 15 years associated with low-risk prostate cancer was

S. Agrawal · S. Yadav · A. Kumar (✉)
Department of Urology, Robotics and Renal Transplant, Safdarjung Hospital and Vardhman Mahavir Medical College, New Delhi, India

N. Gupta (ed.), *A Clinical Guide for Cancer Prostate Patient*,
https://doi.org/10.1007/978-981-97-5357-4_7

found to be 2.7% [3]. Death in prostate cancer patients most commonly occur due to other causes or competing diseases such as cardiovascular disease and rarely due to the prostate cancer itself. Besides, low risk prostate cancer patients tend to have same quality of life as their healthy counterparts, and the cancer per se doesn't affect the quality of life.

7.3 What are the Concerns in Men Diagnosed with Low Risk Prostate Cancer?

Men with low-risk prostate cancer have low risk of dying from the disease, even 15 years after the diagnosis. Besides, whether Gleason score 6 should be called a cancer or not has also been debated. There are some things that men recently diagnosed with low-risk prostate cancer should keep in mind.

- Pure Gleason score 6 (i.e., only Gleason pattern 3) tumors are not capable of causing significant harm by local or systemic spread in the vast majority of cases.
- However, the diagnosis is made on needle biopsy, which may miss higher grade cancer in up to 35% of the cases. Missing higher grade cancer is probably the most important factor why patients with low-risk prostate cancer should be kept under follow-up [4].
- Also, patients with low-risk prostate cancer (pure GS 6 tumor) have a small risk (2–4%) of progression of disease and merits follow-up.

Multiparametric magnetic resonance imaging (MRI) has been shown to reduce the chances of missing higher-grade cancer and is widely used these days.

7.4 What are the Options for Treatment in Men with Low-Risk Prostate Cancer?

The main risk of patients with low-risk disease is over treatment. As the disease has indolent course, a shared decision should be taken after understanding the disease and all the options of treatment. Following are the options for management [1]:

1. **Watchful Waiting**: Watchful waiting refers to conservative management for patients deemed unsuitable for curative treatment from the outset, and patients are clinically 'watched' for the development of local or systemic progression with disease-related complaints. When the disease progress, then they are treated

palliatively according to their symptoms in order to maintain good QoL (Quality of Life). It is preferred in the patients who have co-morbidities, and the life expectancy is less than 10 years. Main aim here is to treat only if symptoms occur as majority of these patients are unlikely to benefit from treatment of the cancer.

2. **Active Surveillance**: Active surveillance (AS) involves close monitoring of the cancer and curative treatment is initiated at the first sign of progression. Close monitoring is done by a predefined schedule. AS is indicated in men who have life expectancy more than 10 years, and this strategy aims to minimize treatment related toxicity while still maintaining high cancer specific survival.

Strict follow up criteria during AS includes:

- Digital rectal examination (at least once yearly)
- Prostate-specific antigen (PSA) (at least once every 6 months)
- Repeated biopsy every 2–3 years
- Magnetic resonance imaging (MRI) and repeat biopsy if PSA is rising (PSA-doubling time <3 years).

- **Radical Prostatectomy**: It involves surgical removal of the entire prostate along with seminal vesicles. It can be done by open surgery or minimally invasive surgery (laparoscopic or robotic assisted). It aims to remove the disease completely from the body, although it carries risk of urinary incontinence and erectile dysfunction
- **Radiotherapy**: It can be given in two ways external beam radiotherapy (EBRT) and brachytherapy. EBRT is not invasive and involves delivering radiation (high energy X-rays) to prostate gland from outside the body. The treatment is given daily for few weeks. Brachytherapy involves placement of radioactive seeds directly in prostate gland, delivering high dose of radiation directly in the prostate gland. Both of this modality carries risk of radiation injury to bowel and bladder
- **Focal Therapy**: Focal therapy involves treating only the cancerous portion of the prostate gland while sparing the surrounding tissue. This can be done using techniques such as cryotherapy or high-intensity focused ultrasound (HIFU). Their benefits are still not clearly defined, as long-term studies are lacking (Table 7.1) compares different options of low risk prostate cancer treatment.

Table 7.1 Comparing different options of treatment for low-risk prostate cancer

Treatment option	Advantages	Disadvantages
Active surveillance	Avoids unnecessary treatment Can delay or avoid potential side effects of treatment	Risk of progression and need for definitive treatment Need for frequent monitoring and biopsies Anxiety about cancer progression
Radical prostatectomy	High cure rate at single admission Clear pathology Long-term follow-up Anxiety free improved quality of life as whole gland is removed	Urinary incontinence (2.5–8%) Erectile dysfunction (30–50%) Shortening of penis by 1–2 cm Complications of surgery Recovery time Anesthesia side effects
Radiotherapy	High cure rate Non-invasive Minimal recovery time Minimal side effects	Multiple visits to hospital for RT Erectile dysfunction (30–40%) Urinary incontinence Radiation cystitis Radiation enteritis and proctitis Secondary cancers Long term follow-up
Focal therapy (focal ablation, HIFU, cryotherapy)	Non-invasive Minimal recovery time Minimal side effects Shorter hospital stay compared to surgery	Limited long-term data Higher risk of needing further treatment Erectile dysfunction Rectal injury Costly Deficient in long term follow up studies

7.5 Which Would Be the Preferred Center of Choice?

The preferred hospital would be a hospital which provides in house integrated care with all departments (urology, radiotherapy, medical oncology) and can help in shared decision making, follow up and further treatment if needed.

7.6 Which Is the Preferred Treatment of Choice?

Although it depends on choice of the patient and shared decision making of the patient party and the treating oncology team, the preferred choice usually is active surveillance. Figure 7.1 represents a flow chart for management of low risk prostate cancer.

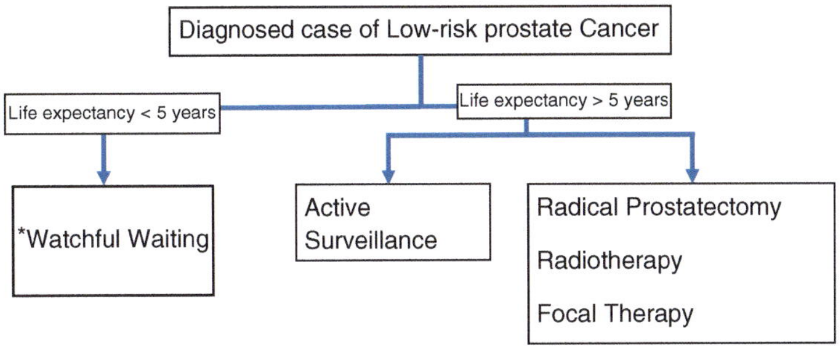

Fig. 7.1 Flow chart

7.7 What are the Key Reasons That a Person with Low Grade Prostate Cancer Choose Active Surveillance?

- It avoids unnecessary over treatment and treatment related side effects.
- Provides an option for delaying the treatment.
- Provides improved quality of life by avoiding physical, mental, and financial stress of undergoing unnecessary treatment. These patients have better urinary, sexual and bowel function compared to patients who underwent immediate management, thus improving quality of life.
- Avoids financial burden associated with treatment.

7.8 What are the Disadvantages of Active Surveillance Over Radical Treatment?

Some disadvantages of AS:

- Need for a prolonged follow up with PSA and biopsies
- There is a risk of progression and eventual treatment
- There is risk of missing higher-grade cancer which was missed on targeted and systematic biopsy
- Anxiety due to prostate cancer

7.9 When to Stop Active Surveillance and Change the Treatment?

When patient is on AS, if rise in PSA level, change is Digital Rectal Examination (DRE) findings and prostate biopsy, leads to detecting a higher-grade prostate cancer, patient should be reclassified and treated according to the higher-grade cancer [1, 5]. Reclassifying to higher grade does not always means further change in treatment, it simply means a patient should understand the disease condition at that point of time and take shared decision tailored for himself.

Around 10% of the patients with active surveillance choose active treatment due to anxiety of being on a continuous surveillance and patient anxiety has been found to be a valid reason for active management [6, 7].

7.10 What are the Symptoms Due to Progression of Prostate Cancer?

These days most of the prostatic malignancies are picked up on screening done during evaluation of lower urinary tract symptoms (LUTS) in ageing male (55–70 years of age). LUTS can be due to various benign causes. Benign causes can co-exist with prostatic malignancy. LUTS (especially voiding symptoms) in low-risk prostate cancer can be managed with alpha blocker (Tamsulosin, Silodosin, Alfuzosin), TURP (transurethral resection of prostate) or completely removing the prostate (Radical Prostatectomy).

Progression of symptoms on AS in low-risk prostate cancer is very rare, and if it happens it appears as local symptoms, in form of difficulty in micturition, hematuria or urinary retention. If the disease spreads outside the prostate to lymph nodes, visceral organs, or bones, then patient can have lower abdominal pain, leg swelling and bone pain.

References

1. EAU Guidelines. Edn. Presented at the EAU Annual Congress Milan 2023. ISBN 978-94-92671-19-6.
2. Albertsen PC, Moore DF, Shih W, Lin Y, Li H, Lu-Yao GL. Impact of comorbidity on survival among men with localized prostate cancer. J Clin Oncol. 2011;29(10):1335–41.
3. Hamdy FC, Donovan JL, Lane JA, Metcalfe C, Davis M, Turner EL, Martin RM, Young GJ, Walsh EI, Bryant RJ, Bollina P. Fifteen-year outcomes after monitoring, surgery, or radiotherapy for prostate cancer. N Engl J Med. 2023;388(17):1547–58.
4. Kulac I, Haffner MC, Yegnasubramanian S, Epstein JI, De Marzo AM. Should Gleason 6 be labeled as cancer? Curr Opin Urol. 2015;25(3):238–45.
5. Willemse PP, Davis NF, Grivas N, Zattoni F, Lardas M, Briers E, Cumberbatch MG, De Santis M, Dell'Oglio P, Donaldson JF, Fossati N. Systematic review of active surveillance for clinically localized prostate cancer to develop recommendations regarding inclusion of intermediate-risk disease, biopsy characteristics at inclusion and monitoring, and surveillance repeat biopsy strategy. Eur Urol. 2022;81(4):337–46.
6. Klotz L, Zhang L, Lam A, Nam R, Mamedov A, Loblaw A. Clinical results of long-term follow-up of a large, active surveillance cohort with localized prostate cancer. J Clin Oncol. 2010;28(1):126–31.
7. Lam TB, MacLennan S, Willemse PP, Mason MD, Plass K, Shepherd R, Baanders R, Bangma CH, Bjartell A, Bossi A, Briers E. EAU-EANM-ESTRO-ESUR-SIOG prostate cancer guideline panel consensus statements for deferred treatment with curative intent for localized prostate cancer from an international collaborative study (DETECTIVE study). Eur Urol. 2019;76(6):790–813.

Chapter 8
Active Surveillance (AS) in Prostate Cancer (P Ca)

Manav Suryavanshi

Active Surveillance is the way .... for Low Grade Prostate Cancer!

- Cancer overdiagnosis and overtreatment is a problem in many organs including Prostate. It is estimated that the "overdiagnosis" rates for prostate if the entire reservoir of disease were being detected would be 87–94% [1].
- The widespread use of PSA has decreased prostate cancer mortality [2]. The disadvantage has been overdiagnosis and overtreatment including many cases of clinically insignificant P. Ca.
- The US Preventive Services Task Force—USPSTF recommendations on the use of PSA for screening, published in 2012, assigned it to level D, mainly based on concerns about overdiagnosis and overtreatment [3].
- In 2018, the recommendation regarding PSA screening was modified to a level C (neutral) recommendation due to the uptake of active surveillance, resulting in a reduction in overtreatment [3].
- To avoid overtreatment patients at low risk of progression must be identified and managed conservatively.
- **Conservative** management of localized prostate cancer can either be **"active surveillance"** or **"watchful waiting."**
- **Active surveillance (AS)** is offered in men who can be offered radical local therapy in case they are reclassified for higher risk of disease progression on regular monitoring.

M. Suryavanshi (✉)
Uro-Oncology and Robotic Surgery, Amrita Institute of Medical Sciences and Research Centre, Faridabad, India
e-mail: manavsuryavanshi@fbd.amrita.edu

N. Gupta (ed.), *A Clinical Guide for Cancer Prostate Patient*,
https://doi.org/10.1007/978-981-97-5357-4_8

- The conceptual basis for active surveillance is that patients are candidates for radical therapy as they have a life expectancy of 5–10 years or more, but do not require it because of the indolent nature of the disease.
- In **Watchful Waiting (WW)** palliative therapy is offered for advanced disease if it manifests clinically. It entails minimal follow up and local therapy is usually not an option in view of age and comorbidities.
- Patients initially managed with **AS** are converted to **WW** if their life expectancy is less than 5–8 years (typically around 80 years of age, or earlier if they have significant comorbidities).

8.1 What are Low and Intermediate Risk Disease in Prostate Cancer (P.Ca.)?

Low Risk

PSA <10 ng/mL and Gleason Score (GS) <7 (ISUP grade 1) and cT1–2a

Intermediate Risk

PSA 10–20 ng/mL or Gleason Score 7 (ISUP grade 2/3) or c T2b

8.1.1 Benefits of AS

- AS came in vogue due to concern about overtreatment, data on favorable molecular genetics of Gleason Pattern 3, long-term data on AS and enhanced ability to identify those patients at risk using modern imaging and molecular biomarkers.
- AS avoids radical local therapy for low risk disease albeit keeping an option of definitive treatment for those patients who are reclassified over time as having higher risk for disease progression. This is usually based on finding higher-grade disease on surveillance biopsy.

8.2 Natural History of Prostate Cancer

- 30% of men in their 30s, 50% in their 50s, and 80% or more in men older than 80 years of age harbor microfocal prostate cancer [4].
- The lifetime risk for death from P. Ca. in North America is 3%.
- The ratio of harboring prostate cancer to death is about 20:1, and the ratio of diagnosis to death is about 6:1.
- Many men diagnosed with localized prostate cancer would remain asymptomatic and undiagnosed without PSA testing and be considered **"clinically insignificant"** prostate cancers.

- The indications for biopsy determine the chances of diagnosing clinically insignificant cancer. Lower the PSA threshold for biopsy, more cores at each biopsy and more biopsy sessions lead to higher likelihood of clinically insignificant cancer being discovered.
- Active surveillance helps manage low-grade prostate cancer conservatively, reserving treatment for those who are upgraded over time.
- Selected patients with intermediate-grade cancer are also candidates for AS.

8.3 Molecular Genetics of Gleason Patterns (3 V/s Patterns 4 and 5)

- Most Gleason pattern 3 cells have relatively normal molecular features and Gleason pattern 4 and 5 cells are altered.
- Due to favorable molecular profile; Gleason pattern 3 usually has benign natural history with no metastatic potential whereas higher-grade patterns can invade and metastasize.
- Spontaneous grade progression from Gleason pattern 3 to pattern 4 or 5 is uncommon and found in 1 to 2% of patients per year.
- Grade progression usually occurs in high-volume Gleason 6 cancers [5].
- 10–20% patients on AS will undergo true, biological grade progression after 10 years.
- AS does not always warrant intervention esp. if the man has a limited life expectancy.
- Men with low-risk prostate cancer managed conservatively report occasional cases of metastatic disease. This happens because the presence of occult higher-grade cancer at the time of diagnosis was missed on the diagnostic biopsy. This is responsible for few patients who progress to metastatic disease despite being diagnosed with low-grade cancer.
- This situation exists in about 30% of men diagnosed on the basis of a 10–12-core systematic biopsy.
- Cancer deaths in men with low-risk cancer managed conservatively are caused by the coexistence of higher-grade cancer missed on the diagnostic biopsy.

8.3.1 AS Outcomes

- P. Ca. mortality is rare in men with low-risk disease managed with Active Surveillance.
- The most common cause of death in these patients is cardiovascular disease.
- With a median follow-up of 9 years, the relative risk for non-prostate-cancer death is 10 times that for prostate cancer mortality [6].

- Hence, most Gleason 6 patients are appropriately managed with surveillance.
- However, AS should be offered selectively to Gleason 7 patients.
- Comparison of radical intervention to various versions of conservative management has shown little or no benefit in 3 RCT's in the low-risk group.
- Approximately 25% of patients are reclassified to higher risk and managed with definitive therapy at 5 years.
- Incorporating MRI and biomarkers improves the already favorable outcome by providing earlier identification of the aggressive cancers.

8.3.2 *Patient Selection*

- Most men with grade group 1 (Gleason 3 + 3 = 6) prostate cancer should be managed with active surveillance.
- Currently, most guidelines make a distinction between very low risk (Epstein criteria, i.e., only one to two positive cores, no core with >50% involvement, and PSA density <0.15) and low risk (any grade group 1 with PSA <10).
- This grading reflects the likelihood of coexistent occult higher-grade cancer, and (possibly) the risk for eventual grade progression.
- High volume Gleason 6 cancer is associated with an increasing risk for coexistent higher-grade cancer. The higher volume manifests itself in a higher PSA density; increased number of systematic cores involved and increased extent of core involvement.
- Thus, a patient with high-volume Gleason 6 cancer is still a surveillance candidate, but requires closer scrutiny to exclude coexistent higher-grade disease, either with mpMRI and targeted biopsy, repeat extended biopsy, and/or a tissue-based genetic assay.

8.3.3 *Watchful Waiting*

- In the 5-year time frame, only patients with very aggressive, high-grade disease are at risk for cancer death [7].
- Therefore in elderly or comorbid patients, where the harm of radical therapy for localized disease is likely to outweigh the benefit; watchful waiting is offered as an appropriate option. This means monitoring patients only with infrequent PSA and clinical assessment, with no biopsies or imaging, and treatment only for symptomatic progression or rarely for metastatic disease.
- The watchful waiting approach is often used for newly diagnosed elderly or comorbid patients with low- to intermediate-risk cancer.

- Patients on long-term active surveillance are "switched" to watchful waiting when their life expectancy becomes less than 5–10 years.
- Some patients with a life expectancy of close to 10 years may still prefer definitive therapy if they are found to have significant disease.
- Patients may be motivated by the desire to avoid a prostate cancer death even if their risk is relatively modest.
- Further, our ability to predict survival based on age and comorbidity is inexact.
- Patient preferences and shared decision making play a role in this population.

8.3.4 *Pitfalls of Active Surveillance*

- Conservative management warrants caution due to mis attribution of grade.
- About 30% of men found to have Gleason 6 (grade group 1) cancer on systematic biopsy are found to have higher-grade disease at radical prostatectomy [8].
- Similarly, about the same proportion will be upgraded if subjected to annual biopsies over 10 years [9].
- Most are upgraded to GG2, and upgrading beyond GG3 is rare.
- The presence of **occult high-grade cancer** can be predicted by a number of clinical factors including PSA density, PSA level, the free versus total ratio, the extent of disease on biopsy (number of cores and extent of core involvement), increasing age and race. In addition, restricted diffusion on MRI is a powerful predictor of high-grade cancer.
- The second phenomenon that puts patients at risk is true biologic grade progression, whereby Gleason pattern 3 progresses to pattern 4 or 5. This is much less common than the misattribution of grade based on systematic biopsies.
- Importantly, many of the cases in which patients have progressed to Gleason 3 + 4 will still have indolent disease as the mortality rate of "low-risk" men managed with surveillance is extremely low.

8.3.5 *Role of mpMRI Prostate on Active Surveillance*

- With the increasing use of MRI and targeted biopsy the rate of undergrading is substantially less with targeted biopsy than with systematic biopsy [10].
- The earlier identification of aggressive high-grade cancer with MRI and targeted biopsies reassures the patient with a negative MRI or targeted biopsy that their risk for harboring high-grade cancer is low.

8.3.6 Outcomes of Grade Group 2 (Gleason 3 + 4) and Grade Group 3 (Gleason 4 + 3) Prostate Cancer on Initial Versus Confirmatory Biopsy

- In patients who undergo prostatectomy while being AS-eligible and their disease gets upgraded on surgical pathology—they have better oncological outcomes than those with initial higher-grade biopsy and concordant surgical pathology who are treated immediately [11].
- Men with low-risk prostate cancer who were initially eligible for AS but presented with more aggressive cancer on confirmatory biopsy are found to have better outcomes than those men who are diagnosed initially with more serious cancer hence suggesting that treatment recommendations should consider the diagnostic pathway.

8.3.7 Subset of Favorable Intermediate (Gleason 3 + 4—Grade Group 2) Prostate Cancer and Emerging Role of AS in This Subset

- AS can be considered in patients with low-volume Grade Group 2 (defined as <3 positive systematic cores and <50% core involvement) or another single element of intermediate-risk disease (i.e. favorable intermediate-risk disease).
- The monitoring schedule should be diligent, given the potential higher risk of progression, development of regional or distant metastases and death of this group compared with patients with low-risk disease.
- During monitoring, if repeat non-MRI-based systematic biopsies reveal >3 positive cores or maximum core involvement >50%/core of Grade Group 2 disease, patients should be reclassified and actively treated.
- Patients with Grade Group 3 disease, or patients with intraductal or cribriform histology, should be excluded [12].

8.4 Other Risk Factors

8.4.1 Race

- Racial differences in prostate cancer incidence and mortality are substantial, and relevant to the surveillance population.
- Although they may have a higher risk for harboring higher-grade cancer, black men are unequivocally candidates for surveillance.

- With the increasing use of MRI to identify those harboring high-grade cancer, they should experience a level of safety comparable to Caucasians.

8.4.2 Family History

- Family history is relevant to an individual's risk for prostate cancer but in most cases it should not drive a decision toward radical treatment.
- The family members of an individual diagnosed with clinically insignificant prostate cancer are more likely to be screened and in many cases also diagnosed with clinically insignificant disease. This phenomenon has resulted in "family history inflation."
- Relatives diagnosed with low-grade disease are surveillance candidates even if a family member has been treated for localized prostate cancer.
- However a history of metastatic disease at a young age (<60) in a first-degree relative is significant and may reflect a germline predisposition to prostate cancer.
- Screening for germline mutations, particularly in genes involved in DNA repair, should be performed in such family members.
- Germline mutations in DNA repair genes (*BRCA1* and *BRCA2*, and *ATM* in particular) predispose to an aggressive phenotype [13].
- By 10 years, 80% of BRCA2 patients were upgraded. BRCA1 and ATM were also associated with increased risk, although not as great [14].
- In men with *BRCA* mutations, definitive therapy should be considered, even if low-grade prostate cancer is present.
- A family history of prostate, breast, or ovarian cancer should invite a screening for DNA repair defects in patients with prostate cancer.
- The role of genetic screening in this population is currently being defined.

8.4.3 Monitoring

- Patients on surveillance should be managed with a combination of 6 monthly PSA, annual digital rectal examination (DRE) and MRI and/or prostate biopsy every 2–5 years after the initial confirmatory biopsy.
- The finding of Gleason score 7 or higher should, in most cases, prompt intervention.
- An increase in volume of Gleason 6 cancer alone should not prompt intervention; but it should prompt a search for higher-grade cancer using MRI or a molecular biomarker.

8.4.4 Magnetic Resonance Imaging

- MpMRI improves the accuracy of targeting significant cancer in men on surveillance compared with systematic biopsies.
- MpMRI is useful before confirmatory biopsy, but systematic biopsies should be performed in addition to targeted biopsies.
- Very low risk patients on active surveillance with a negative MRI can defer systematic biopsies.
- In those at higher risk for upgrading, systematic biopsies are still warranted.

8.4.5 Biomarkers

- Low-risk patients on surveillance have modest benefits with biomarkers, as the risk of clinical progression is 0.2–5% at 15 years. However, selected at risk patients, for higher-grade disease, may benefit from genomic testing.
- There are two potential benefits:
  - reassurance of safety in case of favorable genomic risk score
  - earlier identification of those at risk for disease progression
- Clinical studies have demonstrated a greater utilization of conservative management in those who have access to genomic testing.
- *Genetic tissue assays like Genomic Classifier*—a 22-marker genomic classifier (GC) based on RNA expression can independently predict metastasis after prostatectomy [15].
- The Decipher biopsy test is an accurate predictor for adverse pathology in the favorable intermediate risk population. It can also predict metastasis and disease-specific mortality from diagnostic biopsy specimens of primarily intermediate and high-risk men treated with first line radiotherapy or radical prostatectomy [16].
- In patients with a strong family history of prostate, breast, and/or ovarian cancer, germ line testing for DNA repair defects (BRCA, ATM etc.) is warranted.
- Men with BRCA2 mutations had a 2.8 greater risk of progression to higher-grade cancer.
- Since these mutations increase the rate of acquisition of a mutational burden, the presence of a germ line DNA repair defect, particularly **BRCA2**, is a **contraindication to surveillance** [14].
- Genetic testing of the tissue should be restricted to higher-risk patients. This includes patients with high cancer volume on biopsy, those with high PSA density, or those with Gleason 7 who are surveillance candidates.
- Misattribution of grade (25–30% of cases) is much more common than true biologic progression from Gleason 3–4 or 5 (1–2.3% per year).

- Factors that increase the risk for coexistent higher-grade cancer include PSA density (>0.15), race (African-American), and volume of cancer on biopsy (number of positive cores, extent of core involvement).
- Newly diagnosed men on active surveillance should have a confirmatory biopsy within 1 year, targeting the under sampled areas of the prostate.
- After this, they should have PSA every 6 months and periodic repeat biopsy (every 3–5 years).
- MRI and targeted biopsy along with systematic biopsies are warranted for those at higher risk.

8.5 Lifestyle Management

Counsel AS patients about the benefits of dietary and lifestyle modifications. These also benefit general and cardiovascular health and should include:

1. Regular exercise.
2. Body weight management.
3. Stop smoking.
4. Predominantly vegetable- and grain-based diet; moderate meat esp. including fish and skinless poultry.
5. Limit saturated fat in diet.
6. Take vitamin D, 1000–2000 IU/day esp. during reduced sunlight.
7. Consider a low-dose statin.

8.5.1 Quality of Life (QoL)

- Men on active surveillance have a low rate of anxiety; though anxiety does occur with the diagnosis of prostate cancer but is not increased by the selection of conservative management.
- Improved QoL was consistently seen using validated questionnaires covering voiding and sexual function than those who receive active treatment.
- The QoL is comparable to that of men without prostate cancer [17].

EAU guidelines for Active Surveillance—what are the present recommendations and how to interpret them? [12]

- Active surveillance (AS) follow up protocol—DRE—once yearly, PSA—at least once every 6 months and repeat biopsy every 2–3 years.
- If PSA is rising and PSA-doubling time <3 years—perform MRI prostate and repeat biopsy
- Re-classify patients with low-volume ISUP grade group 2 disease included in AS protocols, if repeat non-MRI-based systematic biopsies performed during

monitoring reveal >3 positive cores or maximum core involvement >50%/core of ISUP 2 disease.
- Any change in treatment to be based on biopsy progression and not on progression on MRI and/or PSA.
- Patients with a PI-RADS 1–2 findings on MRI and a low PSA density (<0.15) may be excluded from repeat biopsy.

Eventually it's a Shared Decision Making between the Patient and the treating Physician where the physician acts as the navigator.

Summary for the management of low-risk disease [12].

8.5.2 Recommendations

- Manage patients with a life expectancy <10 years by watchful waiting.
- Manage patients with a life expectancy >10 years and low-risk disease by Active Surveillance.

8.6 Conclusion

- Active surveillance has been widely adopted as a safe and effective strategy for men with low-risk prostate cancer.
- It is considered the primary treatment for most of these men by numerous national organizations, including the AUA, EAU and ASCO. It helps address overtreatment that accompanies screening and early detection.

References

1. Welch HG, Black WC. Overdiagnosis in cancer. J Natl Cancer Inst. 2010;102(9):605–13.
2. Jemal A, Siegel R, Xu J, et al. Cancer statistics, 2010. CA Cancer J Clin. 2010;60(5):277–300.
3. U.S. Preventive Services Task Force. Final recommendation statement. Prostate cancer screening. 2018. https://www.uspreventiveservicestaskforce.org/Page/Document/RecommendationStatementFinal/prostate-cancer-screening1.
4. Sakr WA, Grignon DJ, Crissman JD, et al. High grade prostatic intraepithelial neoplasia (HGPIN) and prostatic adenocarcinoma between the ages of 20–69: an autopsy study of 249 cases. In Vivo. 1994;8(3):439–43.
5. Inoue LY, Trock BJ, Partin AW, et al. Modeling grade progression in an active surveillance study. Stat Med. 2014;33(6):930–9.
6. Klotz L, Vesprini D, Sethukavalan P, et al. Long-term follow-up of a large active surveillance cohort of patients with prostate cancer. J Clin Oncol. 2015;33(3):272–7.
7. Popiolek M, Rider JR, Andrén O, et al. Natural history of early, localized prostate cancer: a final report from three decades of follow-up. Eur Urol. 2013;63(3):428–35.

8. Epstein JI, Feng Z, Trock BJ, et al. Upgrading and downgrading of prostate cancer from biopsy to radical prostatectomy: incidence and predictive factors using the modified Gleason grading system and factoring in tertiary grades. Eur Urol. 2012;61(5):1019–24.
9. Tosoian JJ, JohnBull E, Trock BJ, et al. Pathological outcomes in men with low risk and very low risk prostate cancer: implications on the practice of active surveillance. J Urol. 2013;190(4):1218–22.
10. Radtke JP, Kuru TH, Bonekamp D, et al. Further reduction of disqualification rates by additional MRI-targeted biopsy with transperineal saturation biopsy compared with standard 12-core systematic biopsies for the selection of prostate cancer patients for active surveillance. Prostate Cancer Prostatic Dis. 2016;19(3):283–91.
11. Perera M, Jibara G, Tin AL, et al. Outcomes of grade group 2 and 3 prostate cancer on initial versus confirmatory biopsy: implications for active surveillance. Eur Urol Focus. 2022;9:662–8. https://doi.org/10.1016/j.euf.2022.12.008.
12. EAU guidelines on prostate cancer. 2023.
13. Taylor RA, Fraser M, Livingstone J, et al. Germline BRCA2 mutations drive prostate cancers with distinct evolutionary trajectories. Nat Commun. 2017;8:13671.
14. Carter HB, Helfand B, Mamawala M, et al. Germline mutations in ATM and BRCA1/2 are associated with grade reclassification in men on active surveillance for prostate cancer. Eur Urol. 2019;75(5):743–9.
15. Klein EA, Haddad Z, Yousefi K, et al. Decipher genomic classifier measured on prostate biopsy predicts metastasis risk. Urology. 2016;90:148–52.
16. Nguyen PL, Haddad Z, Ross AE, et al. Ability of a genomic classifier to predict metastasis and prostate cancer-specific mortality after radiation or surgery based on needle biopsy specimens. Eur Urol. 2017a;72(5):845–52.
17. Venderbos LDF, Aluwini S, Roobol MJ, et al. Long-term follow-up after active surveillance or curative treatment: quality-of-life outcomes of men with low-risk prostate cancer. Qual Life Res. 2017;26(6):1635–45.

Chapter 9
Management of Intermediate Risk Prostate Cancer

A. N. Sachin and Akhbar Salim

9.1 Risk Stratification of Localised Prostate Cancer

- Localised prostate cancer is risk stratified into low risk, intermediate risk and high risk by European Association of Urology (EAU). National Cancer Centre Network (NCCN) has again sub stratified the intermediate risk group into favourable and unfavourable.
- Intermediate risk prostate cancer has attracted a lot of debate in recent times regarding the management.
- This chapter covers pathological and biochemical criteria, sub classifications and management strategies existent for intermediate risk localised prostate cancer.

9.2 Gleason Score

- Gleason grades are based on the glandular pattern of tumour identified at relatively low magnification
- Assigned grade from 1 to 5, with 1 being most differentiated and 5 being most undifferentiated
- On Biopsy *Most common* + *Highest grade* will result in the Gleason score
- Gleason score 6 is the lowest assigned Gleason score on needle biopsy
- Based on the Gleason scores, specimens are allotted to various grade groups ranging from 1 to 5
- **International Society of Urological Pathology 2014 grade group system**:

A. N. Sachin (✉) · A. Salim
Department of Urology, Urologic Oncology, Robotic Surgery and Renal Transplantation, Aster Whitefield Hospital, Bangalore, India

N. Gupta (ed.), *A Clinical Guide for Cancer Prostate Patient*,
https://doi.org/10.1007/978-981-97-5357-4_9

- Grade Group 1 (Gleason Score 3 + 3 = 6): Only individual discrete well-formed glands
- Grade Group 2 (Gleason Score 3 + 4 = 7): Predominantly well-formed glands with lesser component of poorly formed/fused/cribriform glands
- Grade Group 3 (Gleason Score 4 + 3 = 7): Predominantly poorly formed/fused/cribriform glands with lesser component of well-formed glands
- Grade Group 4 (Gleason Score 8)

  Poorly formed/fused/cribriform glands or
  Predominantly well-formed glands and lesser component lacking glands
  Predominantly lacking glands and second most common component of well-formed glands

- Grade Group 5 (Gleason scores 9–10): Lack gland formation (or with necrosis) with or without poorly formed/fused/cribriform glands

9.3 Risk Assessment in Localised Prostate Cancer

- Three components of most acknowledged risk stratification criteria:
  - Clinical stage
  - Biopsy clinical score
  - Pretreatment S. PSA
- EAU/AUA/ASTRO/SIU risk stratifies localised prostate cancer into low risk, intermediate risk and high risk prostate cancer
- Underlining the broad aggressiveness of low risk and intermediate risk prostate cancer, NCCN has proposed a broader risk stratification algorithm
- **Risk stratification and staging work up as advised by NCCN** (Table 9.1)

Table 9.1 NCCN intermediate risk group

NCCN risk group	Features	Imaging
Favourable intermediate	• T2b–T2c OR • Gleason score 3 + 4 = 7/ grade group 2 OR • PSA 10–20 ng/mL AND • Percentage of positive biopsy cores <50%	• Bone imaging: not recommended for staging • Pelvic ± abdominal imaging: recommended if nomogram predicts >10% probability of pelvic lymph node involvement
Unfavourable intermediate	• T2b–T2c OR • Gleason score 3 + 4 = 7/ grade group 2 or Gleason score 4 + 3 = 7/grade group 3 OR • PSA 10–20 ng/mL	• Bone imaging: recommended if T2 and PSA >10 ng/mL • Pelvic ± abdominal imaging: recommended if nomogram predicts >10% probability of pelvic lymph

9.4 Management Strategies for Intermediate Risk Prostate Cancer

- When managed with non-definitive strategies, these cases tend to have 13% 10 years prostate cancer specific mortality (PCSM) and 19.6% at 15 years
- Several treatment options are available for intermediate risk groups
- **Active Surveillance for Favourable Intermediate Group**
  - To avoid unnecessary treatment in clinically localised prostate cancer who do not require immediate treatment
  - Structured surveillance programmes with regular follow up consisting of PSA testing, clinical examination, MP MRI and repeat prostate biopsies, with curative treatment being prompted by predefined thresholds indicative of potentially life threatening disease which is still potentially curable, while considering individual life expectancy
  - Various evidences are available regarding the role of Active surveillance in the management of this subgroup of cases
  - 22% of the patients in ProtecT trial had ISUP grade >1. In ProtecT trial Active monitoring was compared with surgery and EBRT which revealed similar overall survival among 3 groups
  - One Canadian consensus group has also proposed that low volume ISUP Grade 2 (<10% Gleason pattern 4) may be considered for AS
  - American Society of Clinical Oncology (ASCO) also supports these arguments
  - Recent DETECTIVE consensus meeting also support AS for favourable intermediate risk group
  - However, it is clear that the presence of any grade 4 pattern is associated with three fold increased risk of metastasis compared to ISUP grade 1
  - ASCO recommends a confirmation biopsy 6 months after diagnostic biopsy which should be preceded by MP MRI if not done initially
  - PSA should be repeated every 3–6 monthly and DRE annually
  - NCCN, AUA and NICE guidelines also recommends PSA analysis every 6 monthly for patients in Active surveillance
- **Radical Prostatectomy**
  - In SPCG-4 study, death from any cause, death from Prostate Cancer and distant metastasis were significantly lesser in RP arm compared to the watchful waiting arm at 18 years
  - In PIVOT trial RP significantly reduced mortality from all cause compared to Watchful waiting, however it didn't reduce Prostate Cancer associated mortality
  - Recommended treatment modality in patients of this risk group with more than 10 years life expectancy
  - Can be performed by Open, laparoscopic or robot assisted approaches

- Nerve sparing surgery is recommended in patients with low risk of extra capsular disease

- **Lymph Node Dissection**
  - Extended LND provided information regarding staging and prognosis, however it doesn't improve oncological outcomes
  - Extended LND includes removal of the nodes overlying the external iliac artery and vein, nodes within the obturator fossa and nodes medial and lateral to the Internal iliac artery
  - This template allows accurate staging in 94% of the patients
  - As per EAU and NCCN guidelines, extended LND is recommended in all intermediate risk Prostate Ca if the estimated risk exceeds 5% using Briganti nomogram or if it exceeds 7% if modified Briganti nomogram is used

- **Radiation Therapy**
  - Intensity modulated RT with image guided radiotherapy is recommended in all patients planned to undergo EBRT
  - Those who are suitable for Androgen deprivation therapy should be given IMRT with short term ADT (for 4–6 months)
  - Recommended dosage of IMRT is 76–78 Gy
  - Since Prostate Cancer has slow proliferation, hypofractionation is considered more effective than conventional fractions of 1.8–2 Gy
  - Moderate hypofractionation (2.5–3.4 Gy/fx) is recommended over ultra hypofractionation (>3.4 Gy/fn)
  - Those unsuitable for ADT due to comorbidities, IMRT alone or combination of IMRT and brachytherapy is recommended

- **Brachytherapy Monotherapy**
  - Not generally recommended
  - May be offered to favourable intermediate group with <33% biopsy cores involved with cancer and with good IPSS score
  - Supportive evidence for above statement is only available from small series in very experienced centres
  - In selective cases 5 year PSA control rate over 90% are reported
  - Also monotherapy with brachytherapy is associated with minimal high grade genitourinary and GI toxicity

- **Focal Therapy for Favourable Intermediate Group**
  - Focal therapy is advised if disease is confined to one lobe of the prostate. (Details in other chapter)
  - Various ways of focal therapies have been advised for the treatment of prostate cancers, such as

High intensity focused ultrasound
Cryo ablation
Focal laser ablation
Irreversible electroporation
Capsular targeted photo dynamic therapy

– However, none of these has been recommended for definitive treatment of intermediate risk cancer and should only be used under clinical trial settings

Chapter 10
Management of High Risk Locally Advanced Prostate Cancer

T. B. Yuvaraja, Abhinav Veerwal, Akash Shah, and P. Adhikari

High risk locally advanced prostate cancer needs multimodality treatment

10.1 What Is High-Risk Prostate Cancer?

- High-risk prostate cancer is defined as a patient with greater than 50% of failure after primary therapy which includes patients with
  - PSA >20 or,
  - Gleason score 8–10 or,
  - Clinical T stage ≥cT2c [1].

10.2 What Is Locally Advanced Prostate Cancer?

- Locally advanced prostate cancer is a cancer that has broken through the capsule (covering layer) of the prostate gland i.e. the cancer which has breeched the prostatic capsule (T3a), or involved the seminal vesicles (stage T3b), or involvement of adjacent organs (stage cT4) [2].

T. B. Yuvaraja (✉) · A. Veerwal · A. Shah · P. Adhikari
Kokilaben Dhirubhai Ambani Hospital and MRC, Mumbai, India

N. Gupta (ed.), *A Clinical Guide for Cancer Prostate Patient*,
https://doi.org/10.1007/978-981-97-5357-4_10

10.3 What Are the Investigations Done After Prostate Cancer Diagnosis?

- MRI of the prostate is the primary investigation for local staging after diagnosis of prostate cancer [3]. It is employed to improve staging accuracy which can detect extra prostatic extension >1 mm with sensitivity of 70% and specificity of 95% and has 100% specificity for the involvement of seminal vesicles. Further investigation is necessary to rule out metastasis. Bone scan and PSMA PET scan in high risk cases are the preferred investigations for evaluation of prostate cancer metastasis [4].

10.4 What Are the Treatment Options for Locally Advanced Prostate Cancer?

- Treatment for locally advanced prostate cancer is multimodal. There are various treatment options, including surgery (radical prostatectomy i.e. removal of the entire prostate gland), radiotherapy, and hormonal therapy [5].
- One treatment modality is usually not enough and sequencing of treatment should depend on individual case scenario and after discussing the pros and cons of various treatment strategies with patients. Life expectancy of a patient is also an important factor in deciding the line of treatment and primary radical treatment should be offered to patients with life expectancy of more than 5–10 years.
- Study of meta-analysis suggests that the combination therapy improves outcomes in both overall survival and disease specific survival. Combination of RP plus adjuvant RT and hormonal therapy, and RT plus adjuvant hormonal therapy are effective primary treatments, and they have comparable outcomes [6].
- Every treatment modality has its drawbacks in functional outcomes in terms of continence, potency and quality of life (QOL) or treatment related side effects. This complicates appropriate selection of primary treatment even more [7].

10.5 What Is the Role of Surgery as the Modality of Treatment and What Are Its Side-Effects?

- Recent advances in prostate surgery in the form of robot-assisted radical prostatectomy (RARP) have provided a safe, standard, as well as oncologically effective surgical option in the management of patients diagnosed with high-risk localized and/or locally advanced prostate cancer. Radical prostatectomy is performed as a part of multimodal therapy in carefully selected cases. Adjuvant radiotherapy is usually required after surgery due to its locally advanced stage. Outcomes are around 60% of cancer-specific survival at 15 years and 75%

overall survival at 10 years. However, the comparative oncological effectiveness of radical prostatectomy as a part of a multimodal strategy versus upfront radiotherapy with hormonal therapy for locally advanced prostate cancer are currently unknown. A recent retrospective study, evaluating the role of RARP in these patients, found that 188 patients (100%) remained biochemical recurrence-free (BCR) as well as clinical recurrence-free (CR) during the 1- and 3-year follow-up post-RARP. More refinements in RARP, such as super-extended RARP, have been found to be useful in patients with either high-risk localized and/or locally advanced prostate cancer. New extended risk-stratification of radical prostatectomy for non-metastatic high-risk prostate cancer, based on the European Multicenter Prostate Cancer Clinical and Translational (EMPaCT) research database, has been found to be helpful in determining cancer-specific survival (CSS) post-radical prostatectomy in these patients. Complications and side-effects of surgery can be urinary incontinence, impotence, infections, wound-related complications, and others.

- Other advantages of surgery are that in long term it avoids "morbidity of the sitting prostate" like LUTS, hematuria, recurrent urinary obstructions and surgery for bladder neck obstruction. Also, accurate histopathology and T staging is available after surgery to guide further management.

10.6 What Is the Role of Radiotherapy as the Main Modality of Treatment and What Are Its Side-Effects?

- Radiotherapy along with hormonal treatment is another recommended modality of treatment. It has become a contemporary trend in high-risk locally advanced prostate cancer other than surgery. Radiation can be given in form of External Beam Radiotherapy (EBRT) or brachytherapy (low dose or high dose brachytherapy). Radiation may be given upfront as the primary treatment or may be given after surgery as an adjuvant treatment with ADT. (Details in another chapter).
- Urinary difficulties, rectal bleeding/inflammation, and impotency are the concerned side-effects of radiation therapy.

10.7 What Are the Hormonal Therapy and Chemotherapy Approaches for Treatment?

- Androgen Deprivation Therapy (ADT) is available as an adjunct to surgery or radiotherapy. It improves the survival of patients. Chemotherapy is also an option as a part of multimodal treatment in carefully selected patients. (Details in another chapter).

10.8 What Should Be Expected Survival After Getting Treated for Locally Advanced Prostate Carcinoma?

- Surgery or radiotherapy as a part of multimodal therapy gives good results with 75% overall survival rates after 10 years of treatment. Nonetheless, patients have to be adherent to lifelong follow-up with very basic investigations like serum PSA and any further investigations required in the form of MRI or PSMA PET scan as per the clinical judgement of the treating doctor.

10.9 What Are the Chances of Recurrence After Primary Treatment and What Should I Do About It?

- After primary treatment, there are about 25–40% chances of recurrence of the disease in the next 10 years, but all may not have any clinical symptoms. This is why the patients have to be on lifelong follow-up.
- This follow up can be done with periodic serum PSA, followed by MRI or PET scan if required. In case of proven recurrence salvage treatment options are also there like salvage radiotherapy/surgery/hormonal treatment with very good control of the recurrence in terms of overall survival.

10.10 What If the Prostate Cancer Recurs After Taking Primary Therapy?

- Multiple treatment options for prostate cancer recurrence are available so patient need not panic if it recurs after primary treatment.

10.11 Patients Who Have Undergone Radical Prostatectomy

- Patients who have undergone Radical Prostatectomy are further evaluated based on their histopathology reports and follow up serum PSAs and managed accordingly.
- Further in follow up, if there is a rise in serum PSA, patients may need to undergo further investigations in form of PSMA PET scan, bone scan or MRI as determined clinically. Patients may require further treatment in form of ADT, or radiotherapy with ADT which can be adjuvant or early salvage therapy as per clinical decision by the treating doctor. (Details in another chapter on biochemical recurrence).

10.12 Treatment Options for Recurrence for Patients Who Have Undergone Radiotherapy as Primary Treatment

- If prostate cancer recurs after radiotherapy, further treatment is based on whether the recurrence is local or systemic.
- To confirm local recurrence TRUS guided biopsy is done or rise in serum PSA is monitored and after evaluating the patient status and performance score—Patient can undergo surgery or ADT as per the clinical jugdement.

References

1. Thompson I, et al. Guideline for the management of clinically localized prostate cancer: 2007 update. J Urol. 2007;177:2106
2. Greene FL, Page DL, Fleming ID. AJCC cancer staging manual. ed 6. Springer Verlag: New York, NY; 2002:309–16
3. Fulgham PF, Rukstalis DB, Turkbey IB, Rubenstein JN, Taneja S, Carroll PR. AUA policy statement on the use of multiparametric magnetic resonance imaging in the diagnosis, staging and management of prostate cancer. J Urol. 2017;198:832–838.
4. Hofman MS, et al. Prostate-specific membrane antigen PET-CT in patients with high-risk prostate cancer before curative-intent surgery or radiotherapy (proPSMA): a prospective, randomised, multicentre study. Lancet, 2020. 395: 1208.
5. Tilki D, Chen M-H, Wu J, et al. Surgery vs radiotherapy in the management of biopsy Gleason score 9–10 prostate cancer and the risk of mortality. JAMA Oncol. 2019;5:213–220.
6. Lennernäs B, Majumder K, Damber JE, et al. Radical prostatectomy versus high-dose irradiation in localized/locally advanced prostate cancer: a Swedish multicenter randomized trial with patient-reported outcomes. Acta Oncol. 2015;54(6):875–81.
7. Jarosek SL, Virnig BA, Chu H, Elliott SP. Propensity-weighted long-term risk of urinary adverse events after prostate cancer surgery, radiation, or both. Eur Urol. 2015;67:273–80.

Chapter 11
Management of Biochemical Recurrence After Definitive Therapy

Rajeev Kumar

Biochemical recurrence does not mean end of life (NPG)

11.1 What Is Biochemical Recurrence?

- Treatment for localized prostate cancer, with either surgery (radical prostatectomy, RP) or radiation therapy (RT), is expected to cure the patient. After treatment (surgery or RT), patients are advised to get their PSA checked at regular intervals.
- After treatment, PSA is usually tested at 6 weeks and then at 3 monthly intervals for 1–2 years. Thereafter, the test may be repeated less frequently if it is not rising.
- After radical prostatectomy surgery, the PSA should become undetectable on standard PSA tests. However, some new ultrasensitive PSA tests can detect very low values and PSA up to 0.01 ng/mL may be detectable even after surgery. After surgery, PSA above 0.2 ng/mL is called biochemical recurrence (BCR). BCR is confirmed if the PSA is elevated or rising on two or more consecutive tests.
- After RT, PSA is tested at regular intervals and the lowest value that is achieved is called the 'nadir' value and taken as the baseline for future comparison.
- After RT, continuous rise in PSA above the nadir value, particularly if it reaches above 2 ng/mL, is called BCR.
- A rise in PSA may indicate that the disease has come back. Such men need to undergo further evaluation to confirm disease recurrence and the site of recurrence.

R. Kumar (✉)
All India Institute of Medical Sciences, New Delhi, New Delhi, India
e-mail: rajeev.urology@aiims.edu

N. Gupta (ed.), *A Clinical Guide for Cancer Prostate Patient*,
https://doi.org/10.1007/978-981-97-5357-4_11

11.2 What Does BCR Mean?

- BCR is not uncommon and 20–40% of men undergoing treatment for localized disease may have BCR.
- PSA is produced by the prostate gland. If PSA rises (occurrence of BCR), it indicates that prostate cancer may have recurred, producing PSA.
- Although BCR means that the cancer has most likely recurred, it does not mean that all patients have developed cancer spread (metastasis) or will die from their cancer. In many men with BCR, the disease progresses very slowly and they may live a normal or good and long life.
- However, it is important that additional tests be performed to assess where the cancer has recurred and what is the nature of the disease.

11.3 Why Does BCR Occur?

- Surgery or RT can usually cure cancer only if it is confined to the prostate gland.
- When cancer is diagnosed, tests are done to check the extent of disease spread.
- However, CT, MRI and PET scans can detect disease only when it has reached a certain minimum size. It is possible that at the time these tests are done, the disease has already spread to other sites outside the prostate but cannot be detected by the tests.
- BCR can occur if the cancer at these initially undetectable sites starts growing. Additionally, BCR can occur if the disease has spread microscopically outside the prostate gland and is not removed during surgery.

11.4 Can BCR Be Predicted?

- BCR can occur at varying time periods after the initial treatment.
- The risk of BCR varies among patients. This is dependent on the PSA before treatment, type of cancer (grade and genetic variations) and extent of disease (tumor size, stage), among other parameters.
- Certain tests are commercially available to predict the risk of BCR. However, they are used only in selected patients and not routinely indicated in all men. Additional molecular biology tests are also being evaluated for prediction of BCR.
- After surgery, the pathology evaluation of removed prostate gland can also be used to predict BCR. Men who have high grade cancer, cancer that has microscopic spread outside the prostate, cancer at the margin of surgery or cancer that has spread to adjacent structures have a higher risk of BCR.

11.5 What Tests Are Needed in Men with BCR?

- Once BCR is confirmed, tests are done to identify the site of recurrence and to predict the likely future course of disease in such men. The decision to perform these tests depends on the life expectancy, general health of the patient and the rapidity with which the PSA is rising.
- Since many men with BCR do not develop metastasis or die from their disease even after 10 years, it is not compulsory to investigate BCR as soon as it occurs.
- The recurrence may be local (at the site of the prostate), regional (in lymph nodes, around the prostate) or metastatic.
- Most patients will be advised a PSMA-PET scan once PSA reaches 0.5 ng/mL after surgery. At PSA levels below this, this scan does not pick up disease easily.
- If there is no evidence of metastatic disease on PSMA-PET, an MRI of the pelvis may be done, particularly if the pathology report after surgery showed increased risk of residual disease which may not be picked up on PET scan.
- In some men, particularly if the PSA is >10 ng/mL or if PSMA scans are not available, bone scan and CT scan of the abdomen may be obtained.
- For men with BCR after RT, PSMA scan may not be advised till PSA rises further beyond 2 ng/mL.
- After RT, an MRI pelvis or biopsy of the prostate gland may also be advised since the disease may have recurred in the residual prostate gland.
- These tests may be negative in many men with BCR and may have to be repeated if PSA rises further since their likelihood of the tests becoming positive increases as the PSA rises.

11.6 What Is the Treatment for BCR?

- Treatment for BCR depends on many factors. These include:
  - Age and life expectancy of the patient.
  - Primary treatment (surgery or RT).
  - Time since primary treatment.
  - Current PSA and the rate at which it is increasing (PSA doubling time, PSADT).
  - Initial tumor grade (on needle biopsy if RT was performed and specimen biopsy if surgery was performed).
  - Initial tumor stage.
  - Current tumor stage based on the results of the imaging studies (presence or absence of metastatic disease).
- After surgery, men with evidence of metastatic disease on imaging tests are treated as described in subsequent sections of this monograph on metastatic disease. This is called metastatic, hormone sensitive prostate cancer (mHSPC).

- Men where imaging studies do not show metastasis and have a life expectancy of >10 years; RT to the prostate and pelvis, with concomitant androgen deprivation therapy (ADT) is usually prescribed.
- The dose of radiation ranges from 66 to 70 Gy or higher and the duration of ADT may range from 6 months to 3 years.
- Multiple modalities exist for delivering RT. The most common is external beam radiation (EBRT) with various modifications to reduce the dose of radiation delivered to surrounding structures. The RT may be delivered in multiple sittings. There are various protocols including short duration (hypofractionated) or regular duration protocols. In general, short duration protocols may have higher toxicity.
- Proton beam therapy is a newer modality of delivering RT. It requires fewer sessions and is more precise. However, benefits over conventional EBRT have not been clearly established.
- RT is started when PSA reaches 0.5 ng/mL as starting salvage RT early gives better outcomes than starting it late.
- However, RT usually is deferred till the patient has achieved urinary control (continence) and has no other significant complications from the surgery.
- Men with short life expectancy or significant associated illness may not be advised any specific treatment other than observation of PSA and regular follow-up for development of symptoms of further disease progression. In such men, if the disease progresses to metastasis, they are managed as for mHSPC.
- In men who initially received RT, BCR with absence of metastatic disease can be managed with salvage surgery. However, such surgery is more complicated than primary surgery and may have more side-effects.
- In some men with BCR, even if imaging does not show metastasis, adverse features such as short PSADT, high grade or disease in the lymph nodes may be an indication for starting ADT instead of local salvage RT.
- In some men with BCR where imaging does not show metastasis, intermittent hormonal treatment may also be considered. If there is good PSA response with PSA levels becoming undetectable, the interval of hormonal treatment be increased.
- In men with BCR with evidence of disease recurring only in the pelvic lymph nodes, some men may be candidates for surgery for removal of such nodes only, without RT or ADT. However, such treatments need to be individualized to each patient.

11.7 What Are the Outcomes of BCR?

- This is dependent upon many factors which include time since primary treatment, PSADT, current PSA, tumor grade and stage.
- Most men with BCR will not develop metastasis even after 10 years.

- After treatment, survival depends on PSA response and time taken for PSA to start rising again.
- Men with short PSADT (≤12 months) and high Gleason score on biopsy (≥8) or short time to BCR after RT (≤18 months) are at a higher risk of disease progression.
- Some commercial tests such as Oncotype DX Genomic Prostate Score (GPS) and Decipher Genomic Classifier (GC) may help predict survival after BCR after surgery. These tests are not advised regularly.

Suggested Readings

Lowrance WT, Breau RH, Chou R, et al. Advanced prostate cancer: AUA/ASTRO/SUO guideline PART I. J Urol. 2021;205:14.

Lowrance WT, Breau RH, Chou R, et al. Advanced prostate cancer: AUA/ASTRO/SUO guideline PART II. J Urol. 2021;205:22.

Simon NI, Parker C, Hope TA, Paller CJ. Best approaches and updates for prostate cancer biochemical recurrence. Am Soc Clin Oncol Educ Book. 2022;42:1–8. https://doi.org/10.1200/EDBK_351033.

Albisinni S, Diamand R. Understanding biochemical recurrence after radical prostatectomy: trust biology, not a number. Prostate Cancer Prostatic Dis. 2023;26(4):637–8. https://doi.org/10.1038/s41391-023-00646-6.

Lee WR. Rising or persistently elevated serum PSA following radical prostatectomy for prostate cancer: management. https://www.uptodate.com/contents/rising-or-persistently-elevated-serum-psa-following-radical-prostatectomy-for-prostate-cancer-management?topicRef=6950&source=see_link.

Moul JW, Lee WR. Rising serum PSA following local therapy for prostate cancer: definition, natural history, and risk stratification. https://www.uptodate.com/contents/rising-serum-psa-following-local-therapy-for-prostate-cancer-definition-natural-history-and-risk-stratification.

Tourinho-Barbosa R, Srougi V, Nunes-Silva I, Baghdadi M, Rembeyo G, Eiffel SS, Barret E, Rozet F, Galiano M, Cathelineau X, Sanchez-Salas R. Biochemical recurrence after radical prostatectomy: what does it mean? Int Braz J Urol. 2018;44(1):14–21.

Chapter 12
Metastatic Castration Sensitive Prostate Cancer (*mHSPC*)

Puneet Ahluwalia and Narmada Gupta

The treatment of mHSPC is a complex topic, and treatment choice should be individualized (NPG)

12.1 Understanding Metastatic Castration-Sensitive Prostate Cancer

12.1.1 Definition and Overview

- Metastatic castration-sensitive prostate cancer is a stage where cancer has spread beyond the prostate gland to other parts of the body, such as the bones or lymph nodes.
- Prostate cancer has both hormone sensitive and hormone resistant cells. Castration sensitivity means that the hormone sensitive cancer cells still rely on the male hormone testosterone to grow; therefore, hormonal therapy plays a crucial role in treatment.
- Metastatic castration-sensitive prostate cancer is cancer that needs androgens (male hormones) to grow and therefore stops growing when androgens are absent. Many early-stage prostate cancers are androgen-dependent, so decreasing the amount of androgens in the body or blocking their action may be an effective type of therapy [1–3].

P. Ahluwalia
Uro-oncology and Robotic Surgery, Medanta, Gurugram, India

N. Gupta (✉)
Urology, Medanta Kidney and Urology Institute, Medanta—The Medicity, Gurugram, India

N. Gupta (ed.), *A Clinical Guide for Cancer Prostate Patient*,
https://doi.org/10.1007/978-981-97-5357-4_12

12.1.2 Diagnosis

- Diagnosing metastatic castration-sensitive prostate cancer involves a series of tests, which include a physical examination, blood tests (including prostate-specific antigen, or PSA), imaging scans, and sometimes a biopsy of the metastatic sites.
- These tests help determine the extent and aggressiveness of the cancer, which is essential for treatment planning.
- For men with early PSA-only progression after definitive surgical or radiation therapy, traditional imaging tests like bone scans and CT scans of the abdomen and pelvis rarely find metastases that were not already known to be there. Because of this, traditional imaging studies have usually done on people with high serum PSA levels. (e.g., PSA levels of 10 ng/mL or higher).
- PSMA positron emission tomography (PSMA PET) scanning and whole-body magnetic resonance imaging (MRI) have made detecting disease recurrence at lower PSA levels easier.

12.2 Treatment Options for Metastatic Castration-Sensitive Prostate Cancer

12.2.1 Hormonal Therapy

- Hormonal therapy, also known as androgen deprivation therapy (ADT), is the mainstay of treatment for metastatic castration-sensitive prostate cancer.
- It works by reducing testosterone levels in the body, slowing down cancer cell growth.
- ADT can be achieved through medications or surgical removal of the testicles (orchiectomy).
- *ADT drugs include*:
  - GnRH agonists: These momentarily "turn off" testicular androgen production. Cancer cells starve, shrinking the prostate. GnRH agonists are injected every 1, 3, 6 months or given as a 12-month depot. There are different types of GnRH agonists and doses and frequency depend upon the type of preparation used.
  - A "flare" of elevated androgens may occur early in treatment. Due to transient tumor growth, some men may have worsening symptoms, including bone discomfort. At least initially, an "antiandrogen" can be added to the GnRH agonist to avoid this.
  - GnRH antagonist: Degarelix briefly "turns off" the testicles' androgen production. This drug acts faster than a GnRH agonist and may be helpful when there is a high risk of disease progression before treatment.

- Another pill-form GnRH antagonist, Relugolix may be more convenient than either agonist or antagonist shots.
- Combined androgen blockade—Some doctors offer a second drug, called an "antiandrogen," with the GnRH agonist for a short time. Bicalutamide is an antiandrogen. These drugs help start GnRH agonists. GnRH agonists temporarily increase androgen production before stopping it. This increase may temporarily aggravate the malignancy. Antiandrogens prevent flares.
- ADT medication is often favored over surgery. Some countries perform surgical castration more often due to cost and compliance. Surgery can also lower testosterone levels faster than medicines. To seem normal after surgery, men can have prosthetic testicles.

12.2.2 ADT Timing

- It varies on disease stage and risk of progression. Early ADT may be advised in high-risk patients, but if the patient has no symptoms or metastases, ADT may be delayed.
- "Intermittent" ADT may be given. This involves taking breaks from medicine and restarting it. This gives adverse effects time to subside, improving quality of life (see 'adverse effects of ADT' below).
- Metastatic prostate cancer patients who choose the quality of life over survival might consider intermittent ADT.
- Some persons who have received local therapy for early cancer (e.g., prostate surgery) and have had a spike in PSA but no signs of metastatic disease may consider it.

12.2.3 ADT Side Effects

Lower male hormone levels cause the following:

- Low libido and erectile dysfunction
- Hot flashes
- Gynecomastia (breast enlargement in men)
- Muscle loss and fat gain
- "Osteoporosis"—bone thinning and weakening—can raise the risk of bone fractures.
- Type 2 diabetes risk
- A modest chance of developing or worsening coronary heart disease, which can cause heart attacks (disputed)
- These side effects are scary and serious. Not all men have these adverse effects. ADT's adverse effects must be weighed against the possibility of cancer growth

if you don't use it. Most men prefer cancer growth to side effects. Many side effects can be prevented or treated.

12.2.4 Additional Therapies and Combined Therapies

More recently, the creation of more successful systemic therapies has led to their use along with ADT as the first treatment for men with more advanced diseases:

- **Abiraterone/prednisone plus ADT**: Abiraterone works by stopping the intracellular conversion of androgen precursors in the testes, adrenal glands, and prostate tumor tissue. At first, it increased total survival in diseases resistant to castration. More recently, randomized trials showed that combining ADT with abiraterone plus prednisone in patients with high-risk localized nonmetastatic or metastatic castration-sensitive disease lengthens overall survival compared to ADT alone.
- **Enzalutamide or apalutamide + ADT**: Both enzalutamide and apalutamide work as androgen receptor inhibitors because they bind to the place in the androgen receptor where androgen binds. Both drugs help men with prostate cancer that does not respond to hormone therapy. More recently, three randomized studies (ENZAMET [Australian and New Zealand Urogenital and Prostate Cancer studies Group (ANZUP) 1304], TITAN, and ARCHES) showed that ADT plus other treatments helped metastatic CSPC more than ADT alone. Now, both apalutamide and enzalutamide can be used in this situation.
- **Docetaxel plus ADT**: Docetaxel makes men with metastatic prostate cancer immune to castration live longer. Randomized trials have since shown that combining docetaxel with ADT gives patients with high-volume metastatic disease sensitive to castration a clinically meaningful survival benefit.
- **Triplet therapy**: More recently, combinations of ADT with docetaxel plus a second systemic agent (darolutamide or abiraterone) have improved survival over ADT plus docetaxel alone. This can be a preferred approach for people with a high disease burden/high-risk disease who are candidates for docetaxel. (PEACE-1 and ARASENS trials).
- Since there is not enough data to show that one approach is better than another, the specific regimen is usually chosen based on the severity of the disease, comorbidities of the patient and a discussion with the patient about the possible side effects of abiraterone, docetaxel, apalutamide, and enzalutamide, as well as the expected length and cost of treatment [4–11].

12.2.5 Monitoring and Follow-Up

PSA Testing:

- After initiating treatment for metastatic castration-sensitive prostate cancer, regular monitoring is essential.
- This typically involves periodic PSA testing, which measures prostate-specific antigen levels in the blood.
- The PSA test helps assess treatment response and disease progression.

12.2.6 Treatment Response Evaluation

- The response to treatment can vary among individuals.
- It is essential to understand that a rising PSA level does not necessarily mean treatment failure.
- Your healthcare provider will evaluate the trend of PSA levels over time, along with other clinical indicators, to determine the appropriate course of action.

12.2.7 Lifestyle and Supportive Care

Maintaining a Healthy Lifestyle:

- Adopting a healthy lifestyle can complement your treatment and improve your overall well-being.
- This includes regular exercise, a balanced diet, stress management, and avoiding tobacco and excessive alcohol consumption.

Supportive Care:

- Managing the emotional and physical aspects of living with metastatic castration-sensitive prostate cancer is crucial.
- Seek support from your healthcare team, family, friends, or support groups to help you cope with possible challenges.
- Additionally, consider engaging in activities that promote relaxation and self-care.

12.2.8 Genetic Testing

Germline

- Recommended for patients with high-risk, very high-risk, regional, or metastatic prostate cancer and those with a family history of high-risk germline mutations.
- Testing recommended to include *MLH1*, *MSH2*, *MSH6*, and *PMS2* (mismatch repair deficiency [dMMR]/Lynch syndrome) and homologous recombination (HR) genes *BRCA1, BRCA2, ATM, PALB2,* and *CHEK2*.

Tumor

- Testing for microsatellite instability (MSI)/dMMR should be considered for regional/mHSPC and advanced tumors and is recommended for mCRPC.
- Testing for HR gene mutations should be considered for regionally advanced tumors and is recommended for metastatic disease.
- Refer for genetic counselling if alterations are detected.

12.3 Summary

- Metastatic castration-sensitive prostate cancer presents unique challenges, but patients can lead meaningful lives with advances in treatment and supportive care
- Patient preferences, cost, volume of disease, and side effect profiles are important in determining which option is the best for an individual patient
- You can confidently navigate this journey by working closely with your healthcare team, staying informed, and maintaining a positive outlook.
- Remember, you are not alone, and support is available to help you every step of the way.

References

1. Huggins C, Stevens J, Hodges CV. Studies on prostatic cancer: II. The effects of castration on advanced carcinoma of the prostate gland. Arch Surg. 1941;43:209.
2. Cornford P, van den Bergh RCN, Briers E, et al. EAU-EANM-ESTRO-ESUR-SIOG guidelines on prostate cancer. Part II-2023 update: treatment of relapsing and metastatic prostate cancer. Eur Urol. 2023;79:263–82.
3. NCCN Clinical Practice Guidelines in Oncology. https://www.nccn.org/professionals/physician_gls/. Accessed 11 July 2023.
4. Sathianathen NJ, Koschel S, Thangasamy IA, et al. Indirect comparisons of efficacy between combination approaches in metastatic hormone-sensitive prostate cancer: a systematic review and network meta-analysis. Eur Urol. 2020;77:365–72.
5. Chi KN, Chowdhury S, Bjartell A, et al. Apalutamide in patients with metastatic castration-sensitive prostate cancer: final survival analysis of the randomized, double-blind, phase III TITAN study. J Clin Oncol. 2021;39:2294–303.
6. Sweeney CJ, Martin AJ, Stockler MR, et al. Testosterone suppression plus enzalutamide versus testosterone suppression plus standard antiandrogen therapy for metastatic hormone-sensitive prostate cancer (ENZAMET): an international, open-label, randomised, phase 3 trial. Lancet Oncol. 2023;24:323–34.
7. Stenzl A, Dunshee C, De Giorgi U, et al. Effect of enzalutamide plus androgen deprivation therapy on health-related quality of life in patients with metastatic hormone-sensitive prostate cancer: an analysis of the ARCHES randomised, placebo-controlled, phase 3 study. Eur Urol. 2020;78:603–14.
8. Marchioni M, Di Nicola M, Primiceri G, et al. New antiandrogen compounds compared to docetaxel for metastatic hormone sensitive prostate cancer: results from a network meta-analysis. J Urol. 2020;203:751–9.

9. Fizazi K, Foulon S, Carles J, et al. Abiraterone plus prednisone added to androgen deprivation therapy and docetaxel in de novo metastatic castration-sensitive prostate cancer (PEACE-1): a multicentre, open-label, randomised, phase 3 study with a 2 × 2 factorial design. Lancet. 2022;399:1695–707.
10. Wang L, Paller CJ, Hong H, et al. Comparison of systemic treatments for metastatic castration-sensitive prostate cancer: a systematic review and network meta-analysis. JAMA Oncol. 2021;7:412–20.
11. Agarwala A, Bansal S, Gupta NP. Bilateral orchidectomy revisited in management of metastatic hormone-sensitive prostate cancer. Indian J Surg Oncol. 2021;12(3):565–70.

Chapter 13
Management of Castrate Resistant Prostate Cancer (CRPC)

Tanvi Sood and Jyoti Wadhwa

13.1 Introduction

- Most patients with advanced cancer of prostate treated with androgen deprivation therapy (ADT) eventually stop responding to ADT and develop castration resistant prostate cancer (CRPC).
- CRPC is prostate cancer that progresses clinically, radiographically, or biochemically despite castrate levels of serum testosterone (<50 ng/dL).
- The defining criteria for CRPC is as follows [1]:
  - Serum testosterone <50 ng/dL
  - Three consecutive rises of prostate-specific antigen (PSA), 1 week apart, resulting in 25% increases over the nadir, with a PSA >2 ng/mL.
  - Progression or appearance of two or more lesions on bone scan or soft tissue lesions using Response Evaluation Criteria in Solid Tumours (RECIST) and with nodes >2 cm in diameter.
- Clinical outcomes for patients developing advanced disease remain poor and survival outcomes reduce as low as 30% at 5 years as the disease evolves to CRPC stage that is incurable and eventually leads to death.
- Recent advances in treatment of metastatic CRPC offer a lot of hope and are improving the outcome of these patients steadily.

T. Sood · J. Wadhwa (✉)
Department of Medical Oncology and Hematology, Paras Health, Gurugram, India

N. Gupta (ed.), *A Clinical Guide for Cancer Prostate Patient*,
https://doi.org/10.1007/978-981-97-5357-4_13

13.2 Pathogenesis

An understanding of possible mechanisms of resistance to castration strategies helps us develop targeted therapies to improve the outcome of these patients.

- In CRPC, intra tumoral testosterone levels remain high to induce cancer progression. This androgen synthesis inside prostate cancer cells then bypasses the need for circulating androgen, thereby promoting cancer cell growth and survival.
- Prostate cells can produce testosterone from different androgen precursors, including cholesterol and adrenal androgen DHEA.
- Androgen receptor (AR) expression is increased through AR amplification which allows the tumour cells to become hypersensitive to even low testosterone levels.
- Genetic alterations in AR gene or TP53 gene (a tumour suppressor gene) or MYC (an oncogene) confer resistance to ongoing hormonal therapy.

13.3 Stages of CRPC

1. Non metastatic CRPC (CRPC M0) is defined as castrate serum testosterone level (<50 ng/dL) plus one of the following:
   - Local recurrence in prostate bed after surgery or local disease persisting after radiation.
   - No detectable disease at primary site, lymph nodes or bone or organs.
   - Up to 40% men who present with CRPC M0 eventually develop metastasis.
2. Metastatic CRPC (CRPC M1) is defined as castrate serum testosterone (<50 ng/dL) plus the appearance of new lesions.
   - This includes either two or more new bone lesions on bone scan, or a soft tissue lesion using the RECIST criteria.
   - Metastatic prostate cancer may involve the non-regional lymph nodes, bones (most common), liver, lung, or brain.

13.4 CRPC Management

13.4.1 Investigations

- For patients who develop CRPC, blood tests include routine laboratory tests with complete hemogram, liver and kidney function tests, PSA levels (PSA doubling time), and ALP levels.

- PSA doubling time (PSADT) is defined as the number of months it would take for PSA to increase twofold. It is a prognostic factor that assesses risk of metastasis and death from prostate cancer. It also helps identify patients that are more likely to develop clinically significant disease.
- For CRPC M0, PSADT is useful for selecting patients for either observation or second-generation antiandrogen treatments.
- The prostate-specific membrane antigen PET CT is more sensitive imaging test than the conventional bone scan and abdominopelvic CT scan. However imaging needs individualization based of treating doctor's discretion.
- Metastatic lesion biopsy is recommended for CRPC M1 patients.
- MSI/MMR testing should be done on the tissue block. This test helps in ruling out a hereditary cancer syndrome called Lynch syndrome and serves as a biomarker for immunotherapy.
- All patients of metastatic CRPC should be offered germline and tumour mutation testing to check for mutations in homologous recombination repair (HRR) pathways genes (BRCA1, BRCA2, ATM, PALB2, CHEK2, etc.). Approximately 25% patients with CRPC M1 have alterations associated with HRR pathways in the tumour tissue. Another 5–10% account for germline (hereditary) HRR mutations, BRCA2 germline mutation being the most common. This information can be helpful for selecting patients for the use of oral therapies called PARP inhibitors and genetic counselling [2].
- Tumour mutation burden (TMB) status testing with next generation sequencing can be done to offer pembrolizumab in later lines of therapy for high TMB status patients.
- Circulating tumour cells (CTCs) can be detected in blood of these patients and their increasing number correlates with shorter survival. Their serial reduction in numbers may indicate a response to ongoing therapy. CTCs estimation has not yet found a place in routine practice yet [3].

13.4.2 Principles of Treatment

- ADT should not be discontinued in patients with CRPC. In addition, treatment options include sequential usage of secondary hormone agents, chemotherapy, immunotherapy, radionuclides, and targeted therapies.
- Treatment options for patients with metastatic CRPC should be chosen based on prior lines of therapy that can include docetaxel or novel hormone agents (NHA) (abiraterone in the castrate sensitive prostate cancer or enzalutamide, darolutamide or apalutamide in CRPC M0).
- Choice of therapy depends on patient factors such as the clinical condition of the patient, comorbidities and patient preference, disease related factors such as presence of organ involvement, tumour burden, symptoms, and treatment related factors such as prior line of therapy, pre-existing side effects and cost of treatment.

13.4.3 Novel Hormonal Agents in CRPC M0

- It is important to maintain castrate levels of testosterone (<50 ng/dL) for CRPC patients through continued ADT with LHRH agonist or antagonist. The rationale is that the androgen receptor activation and tumour androgen synthesis are possibly the mechanisms of recurrence of prostate cancer in CRPC patients.
- PSA doubling time (PSADT) helps guide patient selection for observation, second generation antiandrogen depending on the risk of progression to clinically significant disease.
- PSADT >10 months—Monitoring can be continued with serial PSA levels and continuation of ADT is adequate for asymptomatic patients.
- PSADT ≤ to 10 months—Second-generation antiandrogen that are novel hormonal agents (NHA) are recommended namely Apalutamide, Darolutamide or Enzalutamide as they have all proven to have survival benefit for this high-risk subgroup of patients.
- Enzalutamide—The addition of Enzalutamide (160 mg/day) to ADT has resulted in longer median overall survival than ADT alone among men with nonmetastatic, castration-resistant prostate cancer and a rapidly rising PSA level (PSADT ≤10 months). It has successfully prolonged the duration of metastasis free survival for such that are at a high risk for clinical progression. Important side effects include risk of falls, hypertension, ischemic heart disease and mental impairment disorders [4].
- Apalutamide—This is an antiandrogen agent that increases the metastasis-free survival and time to symptomatic progression for patients with M0 CRPC with a PSADT ≤10 months. Daily recommended dosage is 240 mg and side effects include rash and hypothyroidism [5].
- Darolutamide—This oral agent is the most recent addition to the treatment options for M0 CRPC with a PSADT ≤10 months. Darolutamide (600 mg twice daily) has shown improvement in metastasis-free survival and overall survival outcomes. Most common side effects include fatigue and rash and has the advantage of no mental impairment disorders and seizures [6].

13.4.4 Secondary Hormonal Therapy for CRPC M1

- Patients with CRPC M1 also require continuation of ADT with an LHRH agonist or antagonist agent to maintain castrate serum levels of testosterone (<50 ng/dL) as a backbone of systemic treatment.
- Secondary hormonal agents include:

  1. Androgen biosynthesis inhibitor—Abiraterone + prednisone [7, 8].
  2. Androgen receptor inhibitor—Enzalutamide.

3. First-generation antiandrogen (flutamide, nilutamide, bicalutamide-preferred agent in view of better toxicity and efficacy profile)—used less often these days.
4. Corticosteroids (hydrocortisone, prednisone, dexamethasone), switching from prednisone to dexamethasone itself may benefit a few patients on progression.
5. Antiandrogen withdrawal—may result in clinical or biochemical response.
6. Ketoconazole (an antifungal agent that inhibits steroid synthesis) plus hydrocortisone.

- Abiraterone (1000 mg) should be taken empty stomach along with oral steroid prednisone 5 mg twice daily at least ½ h before meals. Co-administration of steroids prevents side effects such as hypertension and hypokalaemia resulting from mineralocorticoid excess. Abiraterone can be used as an agent if patient prefers oral therapy, unfit or unwilling for chemotherapy options. Time to initiation of chemotherapy is delayed, and it has shown good benefit for symptom control and overall survival. The important side effects include asthenia, hypertension, hypokalemia, edema, arrhythmias, congestive heart failure and liver dysfunction. A smaller dose of 250 mg administered with a low-fat breakfast is an alternative to the standard dose of 1000 mg taken empty stomach. This lower dose may particularly be useful in patients who live in resource-constraint settings [9].
- Enzalutamide (160 mg daily) is another hormonal agent that has shown benefit in patients with CRPC M1. Important side effects include asthenia, diarrhoea, and seizures [10].
- Both abiraterone and enzalutamide are approved as treatment options prior to usage of docetaxel, however for symptomatic patients with M1 CRPC, initial use of docetaxel is preferred in fit patients for rapid symptom and disease control.
- These oral agents are good alternative treatment options for patients who are not fit for treatment with chemotherapy (docetaxel).
- In all patients of CRPC M1 in the post-docetaxel setting, enzalutamide or abiraterone plus prednisone are good options and preferred before trying other hormonal agents like bicalutamide.
- There is no benefit to combining abiraterone with enzalutamide.

13.5 Non-hormonal Therapy for CRPC M1

13.5.1 Chemotherapy

- Docetaxel with steroid—Three weekly docetaxel regimen with oral steroid for 8–10 cycles is the preferred first-line chemotherapy treatment for symptomatic mCRPC. Low dose 2 weekly docetaxel schedule can be used for elderly patients and have been proven to be efficacious and well tolerated.

- Docetaxel re-challenge can be attempted for patients after progression on a NHA whose cancer had not shown progression on prior docetaxel therapy in the castration-sensitive setting [11].
- Cabazitaxel with steroid—Cabazitaxel can be used as a second line chemotherapy option after docetaxel (dosage 20 mg/m^2 or 25 mg/m^2) for patients with CRPC M1 [12].
- Cabazitaxel 20 mg/m^2 plus carboplatin AUC 4 mg/mL per minute with growth factor support can be considered for fit patients with visceral metastases, low PSA level and bulky disease, high LDH, neuroendocrine histology or unfavourable genomics (defects in at least 2 of PTEN, TP53, and RB1) [13].
- Mitoxantrone with prednisone has showed symptomatic control but no survival benefit.

13.5.1.1 Targeted Therapy—PARP Inhibitors

- Deficiency in DNA repair pathway makes cancer cells susceptible to a group of drugs called PARP inhibitors. Olaparib is a PARP inhibitor that can be used in M1 CRPC patients with a pathogenic germline or somatic mutation in the homologous recombination repair pathway gene (BRCA1, BRCA2, ATM, BARD1, BRIP1, CDK12, CHEK1, CHEK2, FANCL, PALB2, RAD51B, RAD51C, RAD51D, or RAD54L) who have been treated previously with ADT regardless of docetaxel exposure [14].
- Rucaparib is another PARP inhibitor approved for patients with CRPC M1 and a BRCA1 or BRCA2 germline and/or somatic mutation who have been previously treated with ADT and docetaxel [15].
- Combination of Olaparib with abiraterone and prednisone is also approved for patients with metastatic CRPC with BRCA mutation if they have not received abiraterone or enzalutamide earlier.

13.5.1.2 Immunotherapy

- Sipuleucel-T is an immunotherapeutic cancer vaccine that is approved only for asymptomatic patients with CRPC M1 without liver metastasis [16].
- Pembrolizumab (for MSI-High, deficient MMR, or TMB high ≥10 mut/Mb) is recommended for patients with metastatic CRPC who have progressed on both docetaxel and a novel hormone agent [17].

13.5.1.3 Radiopharmaceuticals

- Lutetium-177–PSMA-617 is a treatment option for patients with ≥1 PSMA-positive lesion and/or metastatic disease that is predominately PSMA-positive who have progressed on previous ADT and docetaxel.

- Radium-223 is an option for symptomatic bone metastases without visceral metastasis. (Details in another chapter) (Fig. 13.1).

13.6 Sequencing of Therapies

- The optimal sequence or combination of all these agents is largely unknown. Sequencing decisions are made considering many factors including the distribution, extent and pace of disease, comorbidities, previous treatments (chemotherapy or NHAs), patient preferences and drug availability.
- Typically, Docetaxel is preferred for symptomatic metastatic CRPC patients as the first option.

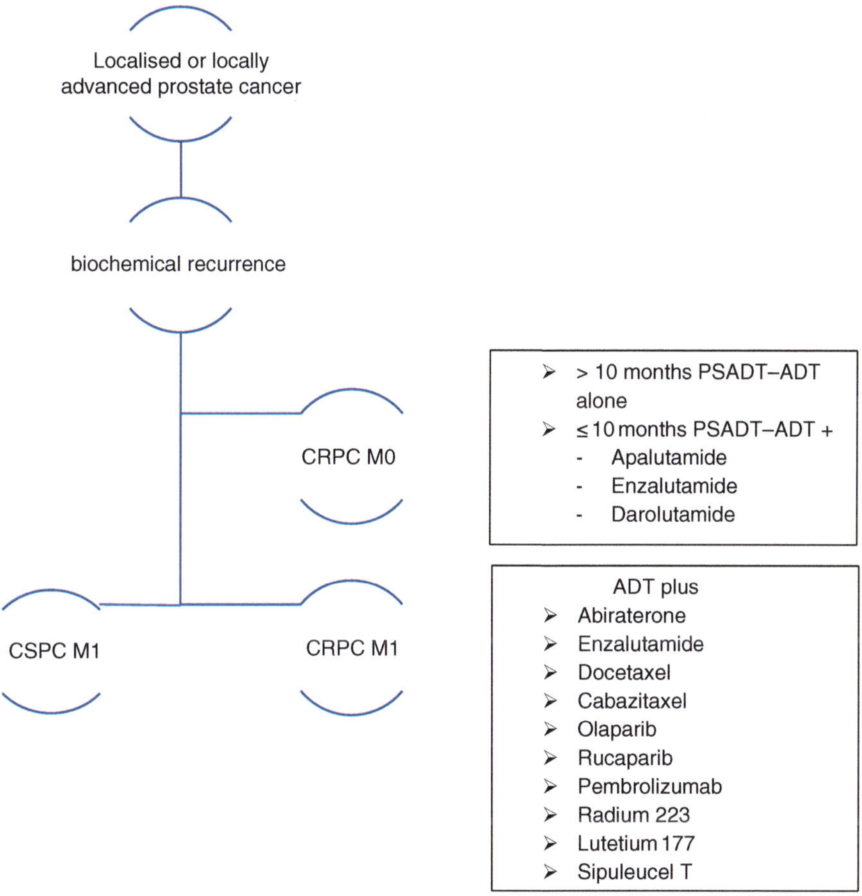

Fig. 13.1 Summary of current systemic treatment options for advanced prostate cancer

- Clinical guidelines for the treatment of CRPC M1 generally recommend considering an alternative mode of action at progression post NHA progression [18].
- Real-world evidence highlights that NHA-to-NHA sequencing is common practice in the treatment of CRPC M1. However, NHA-to-NHA as sequencing deviates from clinical guidelines. There is a lack of evidence to support this approach unless the patient is unfit for chemotherapy. There is reported cross-resistance between abiraterone and enzalutamide that leads to limited clinical benefit with short duration of response [19–21].
- AR-V7 assay is available for predicting abiraterone and enzalutamide resistance in patients with metastatic CRPC, however it is not routinely used in practice for treatment selection.

A lot of advances have resulted from active research in the field of castration resistant prostate cancer, resulting in identification of a few biomarkers to guide treatment decision making, newer targeted therapies, immunotherapy, and radiopharmaceuticals. Outcomes are better now than before. Treatment involves a teamwork of urologist, medical oncologist, radiation oncologist and nuclear medicine expert.

References

1. Scher HI, Halabi S, Tannock I, et al. Design and end points of clinical trials for patients with progressive prostate cancer and castrate levels of testosterone: recommendations of the Prostate Cancer Clinical Trials Working Group. J Clin Oncol. 2008;26:1148–59.
2. Nicolosi P, et al. Prevalence of germline variants in prostate cancer and implications for current genetic testing guidelines. JAMA Oncol. 2019;5:523–8.
3. Goldkorn A, Ely B, Quinn DI, et al. Circulating tumor cell counts are prognostic of overall survival in SWOG S0421: a phase III trial of docetaxel with or without atrasentan for metastatic castration-resistant prostate cancer. J Clin Oncol. 2014;32(11):1136–42.
4. Hussain M, Fizazi K, Saad F, et al. Enzalutamide in men with nonmetastatic, castration-resistant prostate cancer. N Engl J Med. 2018;378:2465–74.
5. Smith MR, Saad F, Chowdhury S, et al. Apalutamide treatment and metastasis-free survival in prostate cancer. N Engl J Med. 2018;378:1408–18.
6. Fizazi K, Shore N, Tammela TL, et al. Darolutamide in nonmetastatic, castration-resistant prostate cancer. N Engl J Med. 2019;380:1235–46.
7. De Bono JS, Logothetis CJ, Molina A, et al. Abiraterone and increased survival in metastatic prostate cancer. N Engl J Med. 2011;364:1995–2005.
8. Ryan CJ, Smith MR, de Bono JS, et al. Abiraterone in metastatic prostate cancer without previous chemotherapy. N Engl J Med. 2013;368:138–48.
9. Patel A, Tannock IF, Srivastava P, et al. Low-dose abiraterone in metastatic prostate cancer: is it practice changing? Facts and Facets. JCO Glob Oncol. 2020;6:382–6.
10. Scher HI, Fizazi K, Saad F, et al. Increased survival with enzalutamide in prostate cancer after chemotherapy. N Engl J Med. 2012;367:1187–97.
11. Petrylak DP, Tangen CM, Hussain MH, et al. Docetaxel and estramustine compared with mitoxantrone and prednisone for advanced refractory prostate cancer. N Engl J Med. 2004;351:1513–20.

12. de Bono JS, Oudard S, Ozguroglu M, et al. Prednisone plus cabazitaxel or mitoxantrone for metastatic castration-resistant prostate cancer progressing after docetaxel treatment: a randomised open-label trial. Lancet. 2010;376:1147–54.
13. Corn PG, Heath EI, Zurita A, et al. Cabazitaxel plus carboplatin for the treatment of men with metastatic castration-resistant prostate cancers: a randomised, open-label, phase 1–2 trial. Lancet Oncol. 2019;20:1432–43.
14. De Bono J, Mateo J, Fizazi K, et al. Olaparib for metastatic castration resistant prostate cancer. N Engl J Med. 2020;382:2091–102.
15. Abida W, Patnaik A, Campbell D, et al. Rucaparib in men with metastatic castration-resistant prostate cancer harboring a BRCA1 or BRCA2 gene alteration. J Clin Oncol. 2020;38:3763.
16. Kantoff PW, Higano CS, Shore ND, et al. Sipuleucel-T immunotherapy for castration-resistant prostate cancer. N Engl J Med. 2010;363:411–22.
17. Marabelle A, Le DT, Ascierto PA, et al. Efficacy of pembrolizumab in patients with non colorectal high microsatellite instability/mismatch repair deficient cancer: results from the phase II KEYNOTE-158 study. J Clin Oncol. 2020;38:1–10.
18. Parker C, et al. Prostate cancer: ESMO Clinical Practice Guidelines for diagnosis, treatment and follow-up. Ann Oncol. 2020;31:1119–34.
19. Noonan KL, North S, Bitting RL, et al. Clinical activity of abiraterone acetate in patients with metastatic castration-resistant prostate cancer progressing after enzalutamide. Ann Oncol. 2013;24:1802–7.
20. Bianchini D, Lorente D, Rodriguez-Vida A, et al. Antitumour activity of enzalutamide (MDV3100) in patients with metastatic castration-resistant prostate cancer (CRPC) pre-treated with docetaxel and abiraterone. Eur J Cancer. 2014;50:78–84.
21. George DJ, et al. Treatment patterns and outcomes in patients with metastatic castration-resistant prostate cancer in a real-world clinical practice setting in the United States. Clin Genitourin Cancer. 2020;18:284–94.

Chapter 14
Genetic Testing in Prostate Cancer

Amlesh Seth and P. Sridhar

14.1 Introduction

Prostate cancer (PCa) ranks as the second most commonly diagnosed cancer following lung cancer, comprising approximately 15% of all cancers diagnoses [1]. As the concept of precision therapy is recently evolving in many cancer therapies, it is not surprising that researchers around the globe are keen on evaluating the same for PCa management.

In recent years, there has been an exponential growth in our understanding of the genetics of PCa, with numerous hereditary syndromes now known to be associated with PCa. This has led to scientists exploring various tests on genetic mutations involved in PCa [2]. There is growing evidence supporting the utility of genetic testing in PCa for both prognostic and therapeutic purposes, particularly in specific patient cohorts. Studies have shown that about 11.8% of men diagnosed with metastatic prostate cancer (mPCa) harbor inherited mutations in DNA repair genes [3]. Likewise, approximately 23% of individuals with castration resistant prostate cancer (CRPC) exhibit aberrations in the deoxyribonucleic acid (DNA) repair pathway as identified through somatic testing [4]. In this chapter, we review the role of genetic tests in PCa and its applicability in Asian men.

A. Seth (✉)
Department of Urology, AIIMS, New Delhi, New Delhi, India

P. Sridhar
Department of Urology, National Cancer Institute, AIIMS, New Delhi, New Delhi, India

N. Gupta (ed.), *A Clinical Guide for Cancer Prostate Patient*,
https://doi.org/10.1007/978-981-97-5357-4_14

14.2 Prostate Cancer: Is It Heritable Cancer?

Numerous hereditary cancer syndromes may be associated with PCa. Hereditary breast and ovarian cancer syndrome follow an autosomal dominant inheritance pattern and is linked to mutations in BRCA1 and 2 genes. Both are tumor suppressor genes located on chromosomes 17 and 13, respectively. These individuals are predisposed to cancers in the breast (in both men and women), ovary, pancreas, prostate and malignant melanoma [5]. Regarding PCa, research indicates that between 11–33% of men with mPCa harbor BRCA1/2 mutations. Furthermore, 5–7% of men diagnosed with early-stage PCa have been identified as carrier of these mutations [6]. It was also found that men with BRCA2 mutations are at increased risk of developing PCa and tend to experience onset at a younger age compared to men with BRCA1 mutations. Furthermore, patients with BRCA2 mutations often present with aggressive disease and exhibit poorer prostate-cancer specific survival rates [7]. Moreover, individuals of Ashkenazi Jewish decent have a greater chance of carrying mutations in either BRCA1 or 2 [8].

Another syndrome associated with PCa is Lynch syndrome, characterized by mutations in DNA mismatch repair genes such as MLH1, MSH2, PMS2, MSH6 and EPCAM. These mutations have been associated with gastro-intestinal malignancies and PCa [9]. Inherited mutations in DNA repair genes occur in 5–10% of localized PCa cases and in 12–17% of metastatic PCa cases [10].

The first hereditary PCa gene identified was the HOXB13 G84E variant. This gene typically interacts with the androgen receptor, influencing the progression of PCa. Carriers of the HOXB13 G84E mutation exhibit a 4- to 4.5-fold heightened likelihood of developing PCa, particularly early-onset cases [11].

14.3 Clinical Implications of Genetic Testing in PCa

14.3.1 Contribution to PCa Screening

The insights on germline testing made the researchers revisiting PCa screening. The IMPACT study evaluated targeted PCa screening utilizing prostate-specific-antigen (PSA) among men carrying germline BRCA1/2 mutations. Their initial results showed that BRCA2 mutation carriers in contrast to noncarriers had higher incidence of PCa, younger age of onset and a greater likelihood of being diagnosed with clinically significant PCa at the time of detection [12]. The IMPACT study group also conducted a prospective evaluation of PSA screening in men with germline mismatch repair pathogenic variants. They found that carriers of MSH2 and MSH6 pathogenic variants had a higher incidence of PCa compared to age-matched individuals without these variants [13].

The recent NCCN guideline proposes contemplating PCa screening for men with a family or personal history indicative of high-risk germline mutations [14]. Specifically, for men with germline *BRCA1/2* mutations, PSA screening should commence at age 40 and continue annually thereafter. This recommendation aligns with those outlined in the NCCN Genetic/Familial High-Risk Assessment: Breast, Ovarian, and Pancreatic guideline, which advocates for initiating PSA screening at age 40 for carriers of BRCA2 mutations and considering the same for carriers of BRCA1 mutations [15, 16].

14.3.2 Role in Active Surveillance

In an active surveillance (AS) cohort, Carter et al. showed that men with localized PCa and BRCA1/2 or ATM mutations were more prone to experiencing grade reclassification compared to those without these mutations [17]. This study highlights the potential role of genetic testing in informing the clinical utility of AS, particularly in BRCA2 carriers with clinically low-risk PCa.

14.3.3 Role in Precision Therapies in Advanced PCa

It is now widely acknowledged that genetic testing can play a vital role in guiding therapeutic strategies for PCa, especially in cases of metastatic or advanced disease. Patients with mPCa who harbor germline and/or somatic mutations in DNA repair genes, such as BRCA1, BRCA2, ATM, PALB2, FANCA, RAD51D, and CHEK2, are more likely to derive benefits from poly (adenosine diphosphate-ribose) polymerase (PARP) inhibitors and platinum-based chemotherapies. This assertion is supported by accumulating evidence demonstrating improved progression-free survival (PFS) and overall survival (OS) across various studies [16].

14.3.3.1 Guidance for Initiating Immunotherapy

In 2017, pembrolizumab gained FDA approval for treating patients with unresectable or metastatic solid tumors exhibiting microsatellite instability-high (MSI-H), or mismatch repair deficient (MMRd) characteristics. It is estimated that around 5% of metastatic CRPC cases possess MSI-H or MMRd genes, rendering them eligible for pembrolizumab therapy [18].

14.3.3.2 Guidance for Initiating PARP Inhibitors

Patients with DNA repair mutations have higher response rates to PARP inhibitors and platinum-based chemotherapy. Currently approved PARP inhibitors for PCa treatment, as well as those under investigation, include olaparib, rucaparib, niraparib, talazoparib, and veliparib.

The PROfound trial enrolled metastatic CRPC patients who had progressed on abiraterone or enzalutamide and harbored mutations in HRR genes identified by tumor sequencing. These patients were divided into two cohorts: cohort A comprised individuals with BRCA1/2 and ATM mutations, while cohort B encompassed patients with mutations in other genes associated with HRR.

The study findings showed that patients in cohort A who received olaparib experienced a significantly longer radiographic PFS compared to those in the control group in cohort A (7.4 vs 3.6 months). Additionally, among evaluable patients, the confirmed objective response rate was significantly higher in the olaparib group than in the control group (33% vs 2%) [19].

Similarly, Rucaparib obtained FDA approval based on findings from the phase 2 TRITON2 trial, which focused on patients with BRCA-mutant metastatic CRPC cases [20]. Further, ongoing investigations are exploring the potential of combining PARP inhibitors with PD-1 inhibitors and other novel hormonal therapies for PCa involving DNA repair defects.

14.3.3.3 Guidance for Initiating Platinum Chemotherapy

Platinum chemotherapy has demonstrated efficacy in treating BRCA1/2 mutated breast and ovarian cancers. A small case series of three patients with mPCa, characterized by biallelic inactivation of BRCA1/2 and progression on initial therapies, exhibited a remarkable response to platinum-based therapy [21].

Another retrospective study investigated the response to platinum chemotherapy in men with metastatic CRPC who had a BRCA2-positive mutation. Among those with the mutation, six of eight individuals (75%) experienced a decline in PSA levels exceeding 50% within 12 weeks, compared to 23 out of 133 men (17%) in controls ($p = 0.001$). Additionally, BRCA2-positive carriers exhibited a median overall survival of 18.9 months from the initiation of chemotherapy, whereas noncarriers had a survival of 9.5 months [22].

In contrast, the prospective PROREPAIR-B study investigated outcomes in metastatic CRPC men who have germline mutations in BRCA1/2, ATM, and PALB2, and were treated with platinum-based therapies. The study found that BRCA2 carriers experienced superior cancer-specific survival (24.0 vs 17.0 months; $p = 0.005$) and a longer progression time to a second therapy (18.9 vs 8.6 months; $p = 0.01$) when initially treated with abiraterone or enzalutamide as opposed to taxanes [23]. Moreover, further prospective studies are needed to assess the effectiveness of first-line AR-targeted therapies compared to taxanes for individuals with mutations in DNA damage repair genes.

14.4 Genetic Testing in Prostate Cancer

The most recent international guidelines and consensus documents offer valuable insights and recommendations regarding the indications and methods for genetic testing in PCa. These resources outline the specific clinical scenarios where the identification of a pathogenic mutation could influence patient management [14]. Depending on the particular indication for genetic testing, the choice of testing method and technique may vary. Factors to consider include whether germline versus somatic testing is appropriate, along with the sample source, such as tumor tissue or blood.

Germline testing typically involves samples obtained from blood, saliva or buccal mucosa whereas somatic testing is performed on tumor tissues obtained from biopsies or resected specimens. In instances where tumor tissue is not readily available, circulating tumor DNA (ctDNA) extracted from blood plasma can be utilized as an alternative sample for somatic testing [24].

14.4.1 Germline Testing

14.4.1.1 Indications of Germline Testing

The recent NCCN guidelines [14] on PCa recommend genetic testing for patients with personal history of PCa diagnosed at any age which ranged between high-risk localised disease to metastatic disease, as well as for men with family history of PCa.

14.4.1.2 Interpretation of Germline Testing

Several commercially available next-generation sequencing panels, such as ProstateNext, Invitae Prostate Cancer Panel, Prostate Gene, OncoGeneDx, are accessible for assessing germline mutations or gene signatures. These panels can aid in guiding therapeutic options, particularly in mPCa exhibiting defective HRR, MSI and high tumor mutational burden.

Germline testing panels may differ in the number of genes being evaluated. These can range from focused panels encompassing approximately 5–6 genes, to cancer-specific panels comprising around 10–15 genes, to comprehensive pan-tumor panels incorporating 80 genes or more. Test results typically return as either pathogenic/likely pathogenic variants, variants of uncertain significance (VUS) and negative findings [16]. Further actions are determined based on the interpretation of these results.

14.4.2 Somatic Testing

The main differentiating point of somatic mutations is unlike germline mutations, somatic aberrations are not inheritable. It is important to note that these acquired mutations arise during the process of cell division. This distinction is crucial to avoid confusion between germline and somatic testing [25].

Several multigene panels containing somatic mutations or specific genomic signatures, such as FoundationOne CDx, FoundationOne Liquid CDx, Guardant360 have been developed. These panels can play a crucial role in guiding precision targeted therapies for mPCa, For example, they can aid in identifying candidates for PARP inhibitors among men with DNA repair mutations or selecting immunotherapy options for those with MSI-H or MMRd alterations, as well as individuals with high tumor mutational burden.

Another important point to remember for both clinicians and patients is that somatic mutations can change over time with ongoing therapies and/or genetic instability [26]. This underscores the importance of a repeat testing over time. Occasionally, somatic testing may uncover an underlying gene mutation that could potentially originate from the germline. In such cases, it is essential to confirm this finding through subsequent germline testing. FDA has recently approved a tissue-based broad companion diagnostic test known as FoundationOne CDx. This is a 324 gene panel test including 14 prostate specific genes which can be clinically utilized in guiding targeted therapies [27].

14.5 Strategies for Genetic Counselling

Figure 14.1 illustrates the accepted model of genetic counselling, which can be divided into pre-test counselling (before genetic testing) and post-test counselling (following the return of test results).

14.5.1 Pre-test Counselling

During the pre-test counselling session, healthcare providers collect information regarding patient's personal medical history, family history, and any previous genetic testing conducted within the patient's family. Genetic counsellors often approach counselling sessions with an educational focus that empowers the patient to make informed decisions about genetic test and to comprehend the options and potential outcomes. The pre-test discussion covers a wide range of topics, including

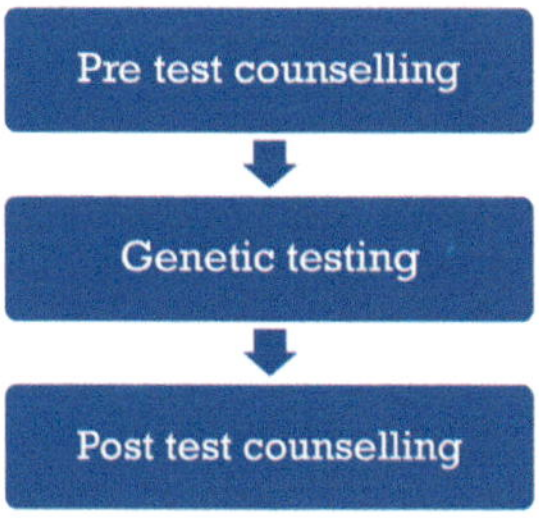

Fig. 14.1 The accepted model of genetic counselling can be divided into pre-test counselling (before genetic testing) and post-test counselling (after test results return)

explanations of hereditary cancer syndromes, assessments of the likelihood of harboring a germline mutation, exploration of different genetic test options to determine the most appropriate one, considerations of potential test results (pathogenic or likely pathogenic variant, VUS, negative), addressing patients and familial concerns and inquiries, and facilitating of financial considerations [8].

14.5.2 Post-test Counselling

During a post-test disclosure session, the patient's genetic test reports are thoroughly reviewed and discussed, accompanied by appropriate recommendations. Beyond addressing emotional and psychosocial needs, the session aims to provide a comprehensive understanding of any medical implications linked to the genetic results for both the patient and their relatives. Discussions concerning medical management are tailored based on the specific outcomes of the genetic testing and the patient's individual family history. Recommendations regarding cancer risk assessment, cascade testing for blood relatives, and the medical management of urological malignancies can all be influenced by the presence of pathogenic or likely pathogenic variants [8].

As such, the genetic counselling process encompasses both pre- and post-test meetings, with a focus on aiding the patient in selecting the most suitable genetic testing approach. Moreover, it considers psychosocial factors that might influence the decision-making process surrounding genetic testing.

14.6 Applicability in Asian Men

Studies have shown that the usual age of PCa onset in Asian men is late as compared with Western populations [28]. Given that the majority of existing data on genetic testing comes from the western world, it remains uncertain whether this is applicable in Asian men with different genetic or racial predispositions. For instance, the mutation in hereditary prostate gene HOXB13 G84E has been linked to early-onset PCa in Western populations. While this specific gene mutation is infrequently identified in Asian populations, studies conducted in China and Japan have reported mutations such as G135E and G132E, respectively [29].

Even though the recent international guidelines suggest genetic testing in PCa in selected population, its applicability in Asian population is still in its infancy. Firstly, data on genetic testing purely from the Asian population is lacking, as there are ethnic disparities with Caucasian men. Second, healthcare facilities with a dedicated genetic counselling team are limited, hindering clinicians from suggesting germline testing even when there is a high probability of a positive germline mutation based on personal and/or family history. Lastly, limited health insurance penetration and consideration of cost involved in germline testing [24].

14.7 Conclusion

Genetic testing is still a new component in the contemporary management of PCa. Various challenges like lack of genetic counsellors, generalizable data and resources for results interpretation and cost to the patient should be addressed for its routine clinical implementation. Clinicians should be cognizant that the detection of a somatic variant in a DNA damage repair gene necessitates subsequent germline testing to ascertain whether an underlying germline pathogenic variant exists. Conversely, when germline testing is conducted, negative results do not exclude the possibility of clinically significant somatic variant. Clinicians may consider initiating these tests simultaneously to optimize the available information for treatment planning.

Acknowledgment None.

Conflicts of InterestNone to declare.

References

1. Ferlay J, Soerjomataram I, Dikshit R, et al. Cancer incidence and mortality worldwide: sources, methods and major patterns in GLOBOCAN 2012: Globocan 2012. Int J Cancer. 2015;136(5):E359–86. https://doi.org/10.1002/ijc.29210.
2. Kafka M, Surcel C, Heidegger I. Recent insights on genetic testing in primary prostate cancer. Mol Diagn Ther. 2021;25(4):425–38. https://doi.org/10.1007/s40291-021-00529-3.
3. Pritchard CC, Mateo J, Walsh MF, et al. Inherited DNA-repair gene mutations in men with metastatic prostate cancer. N Engl J Med. 2016;375(5):443–53. https://doi.org/10.1056/NEJMoa1603144.
4. Robinson D, Van Allen EM, Wu YM, et al. Integrative clinical genomics of advanced prostate cancer. Cell. 2015;161(5):1215–28. https://doi.org/10.1016/j.cell.2015.05.001.
5. El Ansari FZ, Jouali F, Marchoudi N, et al. Screening of BRCA1/2 genes mutations and copy number variations in patients with high risk for hereditary breast and ovarian cancer syndrome (HBOC). BMC Cancer. 2020;20(1):747. https://doi.org/10.1186/s12885-020-07250-0.
6. Messina C, Cattrini C, Soldato D, et al. BRCA mutations in prostate cancer: prognostic and predictive implications. J Oncol. 2020;2020:4986365. https://doi.org/10.1155/2020/4986365.
7. Nombela P, Lozano R, Aytes A, Mateo J, Olmos D, Castro E. BRCA2 and other DDR genes in prostate cancer. Cancers. 2019;11(3):352. https://doi.org/10.3390/cancers11030352.
8. Russo J, Giri VN. Germline testing and genetic counselling in prostate cancer. Nat Rev Urol. 2022;19(6):331–43. https://doi.org/10.1038/s41585-022-00580-7.
9. Ye S, Wang H, He K, et al. Clinical characterization of mismatch repair gene-deficient metastatic castration-resistant prostate cancer. Front Oncol. 2020;10:533282. https://doi.org/10.3389/fonc.2020.533282.
10. Giri VN, Knudsen KE, Kelly WK, et al. Implementation of germline testing for prostate cancer: Philadelphia prostate cancer consensus conference 2019. J Clin Oncol. 2020;38(24):2798–811. https://doi.org/10.1200/JCO.20.00046.
11. Shang Z, Zhu S, Zhang H, Li L, Niu Y. Germline homeobox B13 (HOXB13) G84E mutation and prostate cancer risk in European descendants: a meta-analysis of 24,213 cases and 73, 631 controls. Eur Urol. 2013;64(1):173–6. https://doi.org/10.1016/j.eururo.2013.03.007.

12. Page EC, Bancroft EK, Brook MN, et al. Interim results from the IMPACT study: evidence for prostate-specific antigen screening in BRCA2 mutation carriers. Eur Urol. 2019;76(6):831–42. https://doi.org/10.1016/j.eururo.2019.08.019.
13. Bancroft EK, Page EC, Brook MN, et al. A prospective prostate cancer screening programme for men with pathogenic variants in mismatch repair genes (IMPACT): initial results from an international prospective study. Lancet Oncol. 2021;22(11):1618–31. https://doi.org/10.1016/S1470-2045(21)00522-2.
14. NCCN clinical practice guidelines in oncology: prostate cancer. Version 3. 2023. https://www.nccn.org/professionals/physician_gls/pdf/prostate.pdf. Accessed 8 Aug 2023.
15. NCCN clinical practice guidelines in oncology: genetic/familial high-risk assessment: breast, ovarian, and pancreatic. Version 3. 2023. https://www.nccn.org/professionals/physician_gls/pdf/genetics_bop.pdf. Accessed 8 Aug 2023.
16. Giri VN, Morgan TM, Morris DS, Berchuck JE, Hyatt C, Taplin M. Genetic testing in prostate cancer management: considerations informing primary care. CA Cancer J Clin. 2022;72(4):360–71. https://doi.org/10.3322/caac.21720.
17. Carter HB, Helfand B, Mamawala M, et al. Germline mutations in ATM and BRCA1/2 are associated with grade reclassification in men on active surveillance for prostate cancer. Eur Urol. 2019;75(5):743–9. https://doi.org/10.1016/j.eururo.2018.09.021.
18. Sokolova AO, Obeid EI, Cheng HH. Genetic contribution to metastatic prostate cancer. Urol Clin North Am. 2021;48(3):349–63. https://doi.org/10.1016/j.ucl.2021.03.005.
19. De Bono J, Mateo J, Fizazi K, et al. Olaparib for metastatic castration-resistant prostate cancer. N Engl J Med. 2020;382(22):2091–102. https://doi.org/10.1056/NEJMoa1911440.
20. Wu MS, Goldberg H. Role of rucaparib in the treatment of prostate cancer: clinical perspectives and considerations. Cancer Manag Res. 2022;14:3159–74. https://doi.org/10.2147/CMAR.S353411.
21. Cheng HH, Pritchard CC, Boyd T, Nelson PS, Montgomery B. Biallelic inactivation of BRCA2 in platinum-sensitive metastatic castration-resistant prostate cancer. Eur Urol. 2016;69(6):992–5. https://doi.org/10.1016/j.eururo.2015.11.022.
22. Pomerantz MM, Spisák S, Jia L, et al. The association between germline BRCA2 variants and sensitivity to platinum-based chemotherapy among men with metastatic prostate cancer. Cancer. 2017;123(18):3532–9. https://doi.org/10.1002/cncr.30808.
23. Castro E, Romero-Laorden N, Del Pozo A, et al. PROREPAIR-B: a prospective cohort study of the impact of germline DNA repair mutations on the outcomes of patients with metastatic castration-resistant prostate cancer. J Clin Oncol. 2019;37(6):490–503. https://doi.org/10.1200/JCO.18.00358.
24. Chiu PKF, Lee EKC, Chan MTY, et al. Genetic testing and its clinical application in prostate cancer management: consensus statements from the Hong Kong Urological Association and Kong Society of Uro-Oncology. Front Oncol. 2022;12:962958. https://doi.org/10.3389/fonc.2022.962958.
25. Zhang L, Vijg J. Somatic mutagenesis in mammals and its implications for human disease and aging. Annu Rev Genet. 2018;52:397–419. https://doi.org/10.1146/annurev-genet-120417-031501.
26. Cimadamore A, Lopez-Beltran A, Massari F, et al. Germline and somatic mutations in prostate cancer: focus on defective DNA repair, PARP inhibitors and immunotherapy. Future Oncol. 2020;16(5):75–80. https://doi.org/10.2217/fon-2019-0745.
27. Milbury CA, Creeden J, Yip WK, et al. Clinical and analytical validation of FoundationOne®CDx, a comprehensive genomic profiling assay for solid tumors. PLoS One. 2022;17(3):e0264138. https://doi.org/10.1371/journal.pone.0264138.
28. Zhou CK, Check DP, Lortet-Tieulent J, et al. Prostate cancer incidence in 43 populations worldwide: an analysis of time trends overall and by age group: trends in prostate cancer incidence. Int J Cancer. 2016;138(6):1388–400. https://doi.org/10.1002/ijc.29894.
29. Zhu Y, Wei Y, Zeng H, et al. Inherited mutations in Chinese men with prostate cancer. J Natl Compr Cancer Netw. 2022;20(1):54–62. https://doi.org/10.6004/jnccn.2021.7010.

Chapter 15
Management of Bone Metastasis

Tanvi Sood and Rajeev Sood

Bone management is essential in patients having ADT and bone metastasis

15.1 Introduction

- Bone health and maintenance of bone integrity are important components of comprehensive cancer care in both early and late stages of prostate cancer.
- Metastatic prostate cancer can affect bone health in many ways [1]:
  - Androgen Deprivation Therapy (ADT) induced osteoporosis
  - Osteoblastic metastasis
  - Cancer induced osteoporosis
- Common complications due to bone metastases:
  - Bone pain
  - Vertebral collapse or deformity
  - Pathological fractures
  - Spinal cord compression
- Osteoporosis caused by ADT may also add to fracture risk and should be prevented.
- Indians have low bone mineral density (BMD) as compared to western Caucasians [2]. Osteoporotic fractures in India occur commonly in both sexes and may occur at a younger age than in west.

T. Sood
Medical Oncology, Paras Hospital, Gurugram, Haryana, India

R. Sood (✉)
Punjab Health University, Faridkot, Punjab, India

N. Gupta (ed.), *A Clinical Guide for Cancer Prostate Patient*,
https://doi.org/10.1007/978-981-97-5357-4_15

15.2 Bone Metastasis: Incidence and Pathogenesis

- Hematogenous metastases are present in 35% of patients with prostate cancer, most commonly bone metastasis (90%), lung (46%), liver (25%), pleura (21%) and adrenals (13%) [3].
- In prostate cancer, bone metastasis is usually osteoblastic in nature due to increased bone formation and bone mineral density. However, this regenerated bone is abnormal and easily leads to fractures as it lacks the typical lamellar structure of a normal bone and thus termed "woven bone" or weak bone.

15.3 ADT Related Osteoporosis: Incidence and Pathogenesis

- Both medical and surgical ADT are associated with a high risk of developing osteoporosis and fractures with a 20–50% relative increase in fracture risk [4–6]. Medical ADT is associated with a significantly higher risk of fractures than surgical castration [7].
- Longer treatment duration, advanced age and preexisting comorbidities are also associated with greater fracture risk.
- ADT acts on bone by increasing the bone turnover, decreasing the bone mineral density [8], and increasing the fracture risk in all patients regardless of presence of metastasis.
- Bone mineral density of the hip and spine decreases by approximately 2–3% per year during initial therapy and continues to decline steadily during long-term therapy.

15.4 Management of Bone Metastasis

15.4.1 Bisphosphonates

- Bisphosphonates are bone-targeting agents which act by localizing to bone and inhibiting the activity of osteoclast cells that normally cause osteolysis.
- Commonly used bisphosphonates are as follows (Table 15.1)

15.4.1.1 Zoledronic Acid

- Zoledronic acid (4 mg every 3–4 weeks) is a standard of care to reduce the risk for skeletal related events (SREs) for men with metastatic castration resistant prostate cancer (mCRPC) and bone metastases. It also delays the time to first SRE which can be a pathological fracture, spinal cord compression, surgery or EBRT to bone lesions.

Table 15.1 Dosage of bisphosphonates

Bisphosphonate	Standard dosage (mg)	Route of administration	Duration of infusion	Frequency of administration
Clodronate	1600–3200	Oral	NA	Twice daily
Ibandronate	50	Oral	NA	Daily
Ibandronate	6	Intravenous	1–2 h	Every 3–4 weeks
Pamidronate	90	Intravenous	2 h	Every 3–4 weeks
Zoledronic acid	4	Intravenous	15 min	Every 3–4 weeks

- Use of zoledronic acid in patients with metastatic castration sensitive prostate cancer and bone metastases is not approved for SRE prevention as it has not demonstrated lower risk of SREs in this setting [9].

15.4.1.2 Denosumab

- Denosumab is another bone targeting agent that acts by inhibition of NF-κB ligand (RANKL) receptor to reduce osteoclast function and osteolysis. It has been associated with increased bone-metastasis-free survival.

15.4.1.3 Dosage and Indication

- Zoledronic acid given intravenously every 3–4 weeks or denosumab 120 mg subcutaneously every 4 weeks is recommended to delay disease related SREs in patients with mCRPC and bone metastasis. Both agents reduce SREs at similar rates.
- Reduced frequency of dosing of zoledronic acid and denosumab every 12 weeks is convenient, cost effective and similarly effective to monthly dosing [10].

15.5 Management of ADT Related Osteoporosis

- DEXA Scan—A baseline DEXA scan (dual-energy X-ray absorptiometry) is recommended for all patients on ADT. It is a computer-generated bone mineral density (BMD) scan that assesses percentage of bone mass for subject's age (Z-score) as well comparison to young, healthy adult (T-score).
- WHO BMD criteria for bone health (Table 15.2).
- The National Osteoporosis Foundation guidelines recommend calcium 1000 mg daily and vitamin D3 400–1000 IU daily.
- Additional treatment is recommended for males aged 50 years or more with low bone mass (T-score between −1.0 and −2.5, osteopenia) and a 10-year probability of hip fracture greater than or equal to 3% or a 10-year probability of a major osteoporosis-related fracture greater than or equal to 20% [11].

Table 15.2 DEXA measurement score

	Interpretation of DXA measurement	T score
Normal	BMD more than 1 SD below the young adult female reference mean	≥1
Osteopenia	BMD more than 1 SD but less than 2.5 SD below the young adult mean	<1 and >2.5
Osteoporosis	BMD 2.5 SD or more below the young adult female mean	≤2.5
Severe osteoporosis	BMD 2.5 SD or more below the young female adult mean in the presence of one or more fragility fractures	≤2.5 and clinical fragility fracture

Abbreviations: *WHO* World Health Organization, *DEXA* dual energy X-ray absorptiometry, *BMD* bone mineral density, *SD* standard deviation

15.5.1 Dosage and Indication of Bone Targeted Agents

- For all prostate cancer patients on ADT, osteoporosis related SRE prevention is recommended with denosumab 60 mg subcutaneously once every 6 months or zoledronic acid 5 mg iv once a year, or alendronate 70 mg orally once a week. Follow up DEXA scan should be done after 1 year of starting therapy.

15.5.2 Side Effects and Caution During Treatment with Bone Targeted Therapies

- Baseline dental examination and prophylaxis are recommended before starting any bisphosphonate therapy [12].
- The risk of osteonecrosis of jaw (ONJ) is increased by a history of
  - Trauma
  - Dental surgery
  - Dental infection
  - Intravenous long-term bisphosphonate administration
- Oral hygiene and avoidance of invasive dental surgery during therapy are recommended to reduce the risk of ONJ. If invasive dental surgery cannot be avoided, then therapy with bisphosphonates should be deferred until 4–6 weeks before and after the procedure.
- Bisphosphonates administration may be associated with a self-limited flu-like episode that occurs within the first 24 h of the infusion and can be managed with simple NSAIDs.
- Kidney dysfunction and creatinine level abnormalities can rarely happen with zoledronic acid. Dose reduction is recommended if the creatinine clearance is less than 60 mL/min and completely avoid using zoledronic acid if the creatinine clearance is less than 30 mL/min. However, bisphosphonate-induced renal failure is very rare (0.5%).

- Treatment-related toxicities with zoledronic acid and denosumab are similar and include hypocalcemia (more common with denosumab 13% vs. 6%), temporary bone pains, and osteonecrosis of the jaw (ONJ, 1–2% incidence). Most, but not all, patients who develop ONJ have preexisting dental problems.
- Regular supplementation with calcium and vitamin D are very important in all patients on bisphosphonates to prevent treatment related hypocalcemia.
- Serum calcium and serum creatinine levels should be monitored before starting infusions.

15.6 Management of Complications of Bone Metastasis

15.6.1 Complications

- Bone metastases in advanced Ca prostate are often complicated with severe pain, pathological fractures and sometimes with oncological emergencies like spinal cord compression.
- Painful bone lesions are the most common presentation from bone metastases in prostate cancer.
- Most common sites of fracture include vertebral bodies, pelvic bones, ribs and long bones.

15.6.2 Pain

- The World Health Organization recommends a three-step analgesic ladder approach for palliation of cancer pain.
- For severe bone-related pain, corticosteroids like dexamethasone, methylprednisolone and prednisone are effective in pain control. Oral dexamethasone is commonly preferred because of its high anti-inflammatory action and low mineralocorticoid related side effects.
- Analgesic agents used in the three-step pain palliation approach include:
  - Non-opioid analgesics (acetaminophen, NSAIDs, COX-2 inhibitors)
  - Adjuvants (antidepressants, anticonvulsants, muscle relaxants, alpha-2 adrenergic agonists, NMDA receptor antagonists)
  - Opioids/opioid-like analgesic agents
- Palliative radiation therapy is recommended for managing severe painful lesions specially on weight bearing bones and 80% of patients can expect to achieve pain relief within 2 weeks of therapy.
- Beta-emitting radioisotopes might also provide small benefit with complete reduction in pain over 1–6 months. Strontium-89 and Samarium-153, can partially or completely decrease bone pain in up to 70% of patients of metastatic prostate cancer.

15.6.3 Vertebral Fractures

- Vertebral augmentation comprises of procedures like percutaneous vertebroplasty (bone cementing) and percutaneous kyphoplasty (restoration of vertebral body height) which are effective for pain palliation.

15.6.4 Spinal Cord Compression

- Compression of spinal cord or cauda equina is a disabling complication and a medical oncological emergency that can be assessed by MRI spine.
- Patients should urgently receive a dexamethasone bolus (10 mg intravenously) followed by 4-mg doses every 6 h.
- Most patients need radiation treatment or surgery after assessment by the clinician.

References

1. Gralow JR, Biermann JS, Farooki A, Fornier MN, Gagel RF, Kumar RN, Shapiro CL, Shields A, Smith MR, Srinivas S, Van Poznak CH. NCCN task force report: bone health in cancer care. J Natl Compr Cancer Netw. 2009;7(Suppl 3):S1–32; quiz S33–5.
2. Malhotra N, Mithal A. Osteoporosis in Indians. Indian J Med Res. 2008;127:263–8.
3. Bubendorf L, Schöpfer A, Wagner U, Sauter G, Moch H, Willi N, Gasser TC, Mihatsch MJ. Metastatic patterns of prostate cancer: an autopsy study of 1,589 patients. Hum Pathol. 2000;31(5):578–83.
4. Shahinian VB, Kuo YF, Freeman JL, Goodwin JS. Risk of fracture after androgen deprivation for prostate cancer. N Engl J Med. 2005;352:154–64.
5. Smith MR, Boyce SP, Moyneur E, et al. Risk of clinical fractures after gonadotropin-releasing hormone agonist therapy for prostate cancer. J Urol. 2006;175:136–9; discussion 139.
6. Smith MR, Lee WC, Brandman J, et al. Gonadotropin-releasing hormone agonists and fracture risk: a claims-based cohort study of men with nonmetastatic prostate cancer. J Clin Oncol. 2005;23:7897–903.
7. Sun M, Choueiri TK, Hamnvik OP, et al. Comparison of gonadotropin-releasing hormone agonists and orchiectomy: effects of androgen-deprivation therapy. JAMA Oncol. 2016;2:500–7.
8. Daniell HW, Dunn SR, Ferguson DW, et al. Progressive osteoporosis during androgen deprivation therapy for prostate cancer. J Urol. 2000;163:181–6.
9. Smith MR, Halabi S, Ryan CJ, et al. Randomized controlled trial of early zoledronic acid in men with castration-sensitive prostate cancer and bone metastases: results of CALGB 90202 (alliance). J Clin Oncol. 2014;32:1143–50.
10. Himelstein AL, Foster JC, Khatcheressian JL, et al. Effect of longer interval vs standard dosing of zoledronic acid on skeletal events in patients with bone metastases: a randomized clinical trial. JAMA. 2017;317:48–58.
11. World Health Organization. WHO fracture risk assessment tool.
12. Coleman RE. Risks and benefits of bisphosphonates. Br J Cancer. 2008;98:1736–40.

Chapter 16
Palliative Care for Prostate Cancer

Md. Abdus Salam

Palliative care adds quality life to the remaining years (NPG)

16.1 Introduction

- The aim of treatment of metastatic prostate cancer should be on bringing an adequate control of the disease and it's complications as well as to save the patients from various side effects of the drugs. This principle of management will be achieving a longer progression free survival and improving quality of life through appropriate oncological treatment and palliative care.
- However, in metastatic stages of the disease, when the disease is not responding to any form of treatment, the focus shifts to palliative care. The aims and objectives of palliative care is to improve the quality of life of patients by managing their symptoms along with the patients and the family providing emotional and spiritual support.
- Palliative care is an essential component of the management of metastatic prostate cancer. Palliative care involves a multi-professional approach to managing severe disease with a limited prognosis [1]. There is a lack of specific guidelines in using palliative interventions [2]. Controlling physical, psychological, spiritual, and social symptoms requires those closest to the patient (Fig. 16.1).
- Dame Cicely Saunders launched the contemporary hospice service in 1967. Cicely Saunders developed the first modern hospice and was primarily responsible for establishing the discipline and culture of palliative care. She advocated for appropriate pain management and asked that dying individuals be treated

M. A. Salam (✉)
Department of Urology, Bangabandhu Sheikh Mujib Medical University, Dhaka, Bangladesh

Urology and Transplantation Foundation of Bangladesh, Dhaka, Bangladesh; http://www.Utfbbd.com

N. Gupta (ed.), *A Clinical Guide for Cancer Prostate Patient*,
https://doi.org/10.1007/978-981-97-5357-4_16

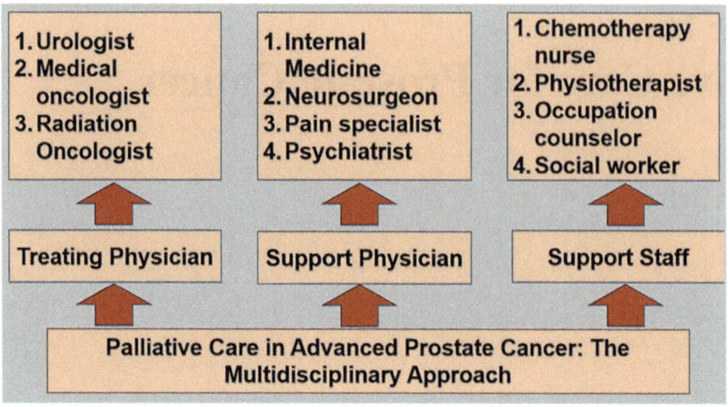

Fig. 16.1 Multidisciplinary approach to the management of advanced and metastatic prostate cancer. (Illustration source: Author)

with dignity, compassion, and respect, as well as rigorous scientific technique when evaluating remedies. She overturned the conventional attitude that patients should be treated, that those who couldn't be cured were failures, and that lying to them about their prognosis was acceptable, if not desirable [3].

- Palliative care should be provided to all patients with life-limiting illnesses, regardless of diagnosis. Palliative care has traditionally been focused on the final weeks and months of life, taking over when active therapy fails. The palliative care team should work seamlessly with all cancer treatment providers to ensure the highest possible quality of life for the patient and their family.

16.2 Different Management Strategies

- **Symptom management**: Advanced prostate cancer symptoms can be quite disabling and negatively impact the patient's quality of life. Palliative care aims to manage these symptoms, which may include pain, fatigue, nausea, vomiting, urinary and bowel problems, and sexual dysfunction. Medications, such as opioids, antiemetics, laxatives, and hormone therapy, are used to manage these symptoms.
- **Emotional and spiritual support**: Palliative care also provides emotional and spiritual support to patients and their families. This includes counselling, support groups, and religious services. The emotional and spiritual needs of patients are often overlooked but are critical in improving their quality of life [4].
- **Advance care planning**: Advance care planning involves discussions between patients, their families, and healthcare providers about the patient's end-of-life care preferences. This includes decisions about resuscitation, life-sustaining treatments, and hospice care. Advance care planning ensures that the patient's wishes are respected and followed.

16.2.1 Hospice Care

Hospice care is provided to patients with advanced cancer who have a life expectancy of 6 months or less. It focuses on managing the patient's symptoms and providing emotional and spiritual support to the patient and their family. Hospice care can be provided in the patient's home, a hospice facility, or a hospital. Hospice care for advanced prostate cancer may include the following services:

- **Pain and other symptoms management**: Almost every advanced prostate cancer patient suffers from pain. In addition to pain, symptoms such as fatigue, nausea, vomiting, urinary and bowel problems, and sexual dysfunction are also common in these patients. Hospice care providers focus on managing pain effectively to ensure that the patient is as comfortable as possible. This may involve the use of medications such as opioids, which are powerful pain relievers.
- **Assistance with activities of daily living**: As the disease progresses, patients may find it difficult to perform activities of daily living such as bathing, dressing, and eating. Hospice care providers may assist with these activities to ensure that patients are comfortable and have the best possible quality of life.
- **Medical equipment and supplies**: Hospice care providers may provide medical equipment and supplies such as hospital beds, wheelchairs, and oxygen to manage the patient's symptoms effectively.
- **Respite care**: Caregivers play a critical role in the care of patients with advanced prostate cancer. However, caregiving can be physically and emotionally exhausting. Hospice care providers may offer respite care to give caregivers a break from their responsibilities.
- In conclusion, hospice care is a specialized type of palliative care that focuses on managing the symptoms of advanced prostate cancer and providing emotional and spiritual support to patients and their families during the end-of-life stage. Hospice care providers offer a various services like pain and symptom management, emotional and spiritual support, helping with activities of daily living, providing medical equipment's and provide support for caregivers. By providing comprehensive care, hospice care can improve the quality of life of patients with terminal stage prostate cancer and their families during a difficult time.

16.2.2 Pain Control

- Most frequent symptom with advanced prostate cancer is pain. This can have a major negative influence on one's quality of life by making it difficult to eat, sleep, and socialise [5]. Cancer patients may require complicated and challenging pain management.
- Dame Cicely Saunders coined the phrase "Total Pain" to characterise severe suffering on a physical, psychological, spiritual, and social level [6]. While not every patient will encounter every one of these issues, treating one without the

other will unavoidably not work in cases where both are present. For instance, it would not be ideal to try to cure physical pain while there also psychological distress.

- On the other hand, it will be very challenging to cure psychological symptoms without also treating physical discomfort. In order to regulate pain in patients who are terminally ill, somatic and psychological care should be implemented simultaneously using a multidisciplinary approach [7] (Fig. 16.2).
- **Over-the-counter pain relievers**: For mild and moderate levels of pain, pain relievers that don't require a prescription may help. Examples include aspirin, Paracetamol/acetaminophen and ibuprofen.
- **Medications derived from opium (opioids)**: Opioids are prescription medications used to treat moderate to severe pain. Examples of opioids include morphine (Kadian, MS Contin, others) and oxycodone (Oxycontin, Roxicodone, others).
- Some opioids are short-acting medicines, so pain relief comes quickly but you may need to take them more often. Other opioid drugs are long-acting medicines, so pain relief takes longer but the medicine doesn't need to be taken as often. Sometimes short-acting and long-acting opioids are used together.
- **Other prescription medicines**: Other types of medicine can help relieve pain, including antidepressants, anti-seizure drugs and steroids.
- **Nerve blocks**: A nerve block procedure can be used to stop pain signals from being sent to the brain. In this procedure, a numbing medicine is injected around or into a nerve.
- **Integrative therapies**: Some people find some pain relief through acupuncture, massage, physical therapy, relaxation exercises, meditation and hypnosis.
- **Music therapy**: Many National Cancer Institute-designated cancer centre offer music therapy as an integrative treatment for cancer. There have been a surprising number of studies done to examine potential benefits of music for cancer patients. It would seem we're guessing intuitively that music has a role [8]. A decrease in pain was noted in some of the studies mentioned, but the effect of

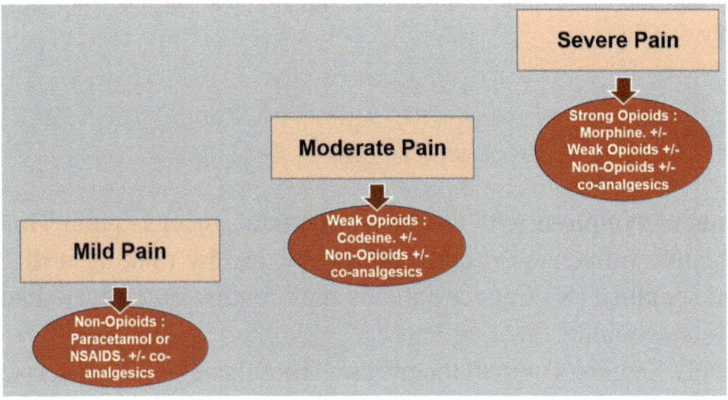

Fig. 16.2 Algorithm of pain management. (Illustration source: Author)

music on pain was studied specifically among people who were undergoing interventions [9].
- **Acupuncture**: Acupuncture may represent a potentially valuable adjunct to existing strategies for pain relief and is known to be relatively free of harmful side-effects [10–12]; however, although acupuncture is widely used in palliative care settings [13] for all types of cancer pain the evidence-base is sparse and inconclusive [14]. Furthermore, safety guidelines for the use of acupuncture treatment in cancer patients exist [15], yet there have been no studies specifically looking at adverse events in patients with bony metastases and no high-quality randomized controlled trials have been conducted to investigate the effects of acupuncture on advanced PC.

16.2.3 Cord Compression

- Spinal cord compression affects 1–12% of patients with metastatic prostate cancer and is an oncological emergency [16]. Prevention of cord compression is important and clinicians should have a high index of suspicion. Disability arising from delay is associated with decreased survival and any function lost before treatment is rarely regained [17]. For this reason, speed is of the essence and investigation and management should not wait for classical signs to develop.
- Treatment should initially be with high dose steroids followed by early radiotherapy or surgery depending on performance status and prognosis [18] (Fig. 16.3).
- It has been demonstrated that decompressive surgery combined postoperative radiotherapy is more successful than radiotherapy alone. However, many people won't be a good fit for this course of care. Overall, skeletal-related-events are best predicted by bone pain [19].

16.3 Preventing and Managing Androgen Deprivation Therapy's Adverse Effects on Prostate Cancer

- Prostate cancer is the second greatest cause of mortality among men from malignancy. Metastatic prostate cancer has a median progression-free survival and overall survival of 18–20 and 24–36 months, respectively. Various modes of androgen suppression therapy are the mainstay treatment as prostate cancer cells are dependent on androgen for growth and differentiation.
- Androgen suppression therapy can have significant and long-lasting results, but it does not provide a complete cure. Additionally, it can cause substantial and annoying side effects. These consequences may have a major impact on a range of aspects of daily life and may further reduce health-related quality of life.

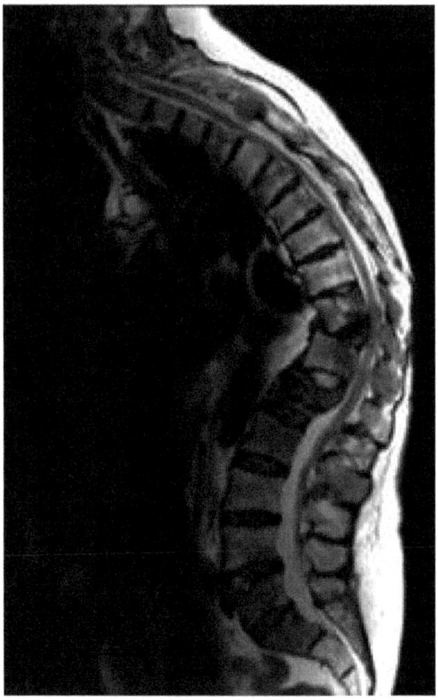

Fig. 16.3 Metastatic lesion in spine can cause cord compression. (Source: Personal collection)

16.3.1 Hot Flashes

- LHRH agonists are widely known for causing a spike in serum testosterone levels during the first week of medication due to the first activation of LHRH receptors, known as the "flare" phenomenon. The flare phenomenon was thought to be the cause of considerable side effects in patients with high metastatic burden.
- A testosterone surge can cause acute clinical progression, but the incidence and severity vary significantly. The clinical ramifications of the flare phenomenon are thought to be minimised by the use of antiandrogens. Antiandrogens reduce the stimulatory action of the testosterone surge at the androgen receptor level. Antiandrogens are added during the first month of LHRH therapy to prevent flare responses in many patients but the there is no specific recommendations regarding this practice.
- Huggins and Hodges reported hot flashes as a common adverse effect of ADT in 1941. Hot flashes are caused by inappropriate activation of thermoregulatory areas in the hypothalamus, which results in peripheral vasodilation [20]. Patients describe hot flushes as excessive warmth followed by cooling, skin flushing, perspiration, and shivers in the top half of the body, particularly the neck and face. Anxiety and palpitations are common symptoms. Hot flashes can last anywhere from a few seconds up to 20 min. Maximum people suffer minimal side effects from hot flashes on rare occasions but in few patients, numerous episodes of hot flashes occur daily [21].

- Medication options for hot flashes include oestrogens, antidepressants, anticonvulsants, and acupuncture. According to a recent study, diethyl stilbestrol and cyproterone acetate had the most efficacy, with up to a 75% reduction of the episodes of hot flushes, however this improvement comes at the expense of irritating side effects [21]. However, because cyproterone acetate is an approved prostate cancer treatment, its usage may conflict with hormonal therapy, and medroxyprogesterone may be the preferred treatment for hot flushes in males undergoing androgen suppression for prostate cancer.
- DES (00.5–1 mg/day), an oestrogen derivative, can lessen hot flashes in 75–90% of males, although there is significant risk of painful gynecomastia. Megestrol acetate, a progesterone derivative, effectively reduced heat flashes by 85% when compared to placebo. To achieve maximum symptomatic reduction, a 2–3-week course of therapy is recommended, with relief lasting several weeks following completion. The initial enthusiasm for megestrol acetate has been discouraged by reports of serum PSA levels rising and then dropping following cessation [22]. Therefore, PSA levels should be continually checked while on treatment.
- Newer antidepressants, including SNRIs (Venlafaxine 37.5 mg/day) and SSRIs (Paroxetine 10 mg/day), may reduce the frequency and severity of hot flashes by increasing serotonin levels and altering the neurotransmitter milieu in the thermoregulatory centre [21, 22]. Both SSRIs and SNRIs are likely to have a mild effect on hot flushes in metastatic prostate cancer patients while on ADT; however, there are no RCT data [23].

16.3.2 Sexual Dysfunction

- ADT can lower serum testosterone levels, which may affect sexual desire and erectile function owing to androgen deficiency. The two most important predictors of ED following ADT were age above 70 years and the presence of diabetes mellitus [24]. Undoubtedly, ED can have a significant impact on the self-perception, self-confidence and quality of life of younger, sexually active men, especially when combined with ADT's deleterious effects on muscle and fat distribution [25].
- Patients with serious erectile dysfunction may be able to get a variety of treatments, depending on the degree of their condition. Every treatment choice will take the patient's preferences into consideration and may have risks and advantages associated with it.

16.3.2.1 Level I Treatment: Oral Medications

- For many men, oral phosphodiesterase inhibitors are an effective treatment for erectile dysfunction. They include Sildenafil, Vardenafil, Tadalafil, Avanafil.

- Nitric oxide, a substance the body naturally makes that relaxes penile muscles, is enhanced by all four drugs. This promotes blood flow and makes it possible to respond to sexual stimulation with an erection. Erection won't happen on its own after taking one of these pills. Nitric oxide is released from the penile nerves only when sexual stimulation is received. For some people, these drugs increase that signal, restoring normal penile function.
- The dosage, duration of action, and adverse effects of the drugs differ. Flushing, nasal congestion, headaches, altered vision, backaches, and upset stomach are among the possible adverse effects. Patients who use nitrate drugs for angina (example, nitroglycerin, isosorbide mono and dinitrate), have other heart disease or heart failure, or have severely low blood pressure (hypotension) should exercise caution when taking this medication. Getting a doctor's advice before using these medications is crucial.

Vacuum Erection Devices: Battery-Operated Penis Pump

- **Penile pumps**: If taking medication alone isn't working or suitable in this situation, a professionally made penile pump may be suggested in addition to the medication. A hand-operated or battery-operated hollow tube with a pump is called a penis pump (vacuum erection gadget). Once the tube is positioned over the penis, the air inside is drawn out by turning on the pump. As a result, a vacuum is created, drawing blood into your penis. To keep the blood in and maintain firmness after the erection, a tension band is wrapped around the base of the penis. Next, take the hoover device out. After sex, the tension ring is taken off. One potential side effect is penile bruising, and ejaculation will be limited by the band. It's possible that a particular model is more comfortable and functional.

16.3.2.2 Level II Treatment: Alprostadil Self-Injection

- Using this technique, inject alprostadil into the side or base of the penis using a tiny needle. Sometimes, drugs that are typically prescribed for different ailments are used alone or in combination for penile injections. Alprostadil, papaverine and phentolamine are the examples. These combination medications are also known as Bimix and Trimix, if two and three drugs are included, respectively.
- Each dose of injection dose is intended for around 1-h erection. The injection site normally hurts very little because of the ultrafine needle that is utilised. Mild injection-related bleeding, priapism, or an extended erection are possible side effects. Rarely, fibrous tissue growth at the injection site may also occur. After an injection, if a priapism appears an hour later, visit the emergency room of the hospital. A straightforward urological procedure to extract the collected blood and irrigate the corpora using ephedrine or phenylephrine.
- **Alprostadil urethral suppository**: A small alprostadil suppository is inserted into the penile urethra as part of alprostadil (Muse) intraurethral therapy. It is

helpful to put the suppository into the urethra using a specific applicator. When successful, the erection typically lasts between 30 and 60 min and begins within 10 min. A burning sensation in the penis, slight urethral haemorrhage, and the development of fibrous tissue inside the penis are possible adverse effects.

- **Testosterone replacement**: Low levels of the hormone testosterone may exacerbate erectile dysfunction in some persons. In this instance, testosterone replacement treatment may be suggested as a first line of treatment or in conjunction with other forms of treatment.

16.3.2.3 Level III Treatment

- **Penile implants**: Devices are surgically inserted into both of the penis' corpora during this treatment. Either inflated or pliable (bendable) rods make up these implants. One can freely choose how long and how long an erection lasts with inflatable gadgets. Your penis remains flexible yet firm thanks to the pliable rods.
- Prior to recommending penile implants, other approaches are typically explored. Implants can be highly successful in the patients with failed conservative management. Similar to every surgical procedure, there is a chance of complications including infection. If you are suffering from a urinary tract infection, you should not have penile implant surgery.

16.3.2.4 Exercise

- Exercise, particularly moderate-to-intense aerobic activity, has been shown in recent trials to improve erectile dysfunction. Regular exercise, even if it's not as intense, may help lower the chance of erectile dysfunction. Increasing your exercise level may help you lower your risk even further.

16.3.2.5 Psychological Counselling

- If stress, anxiety, or depression are the primary reasons of erectile dysfunction, or if the condition itself results in stress and interpersonal conflict, psychological counselling may be necessary. It may be advantageous for a couple to receive counselling from a psychologist or counsellor in addition to their spouse.
- While pharmacologic and mechanical techniques can restore erections, Patients' libido might decline as a result of treatment, limiting willingness to undertake further treatment. Recently, the judicious use of oestrogens has been advocated to improve sexual interest and ED in these men [26].

16.3.3 Management for Reducing Bone Loss and Skeletal-Related Episodes

- In any case a fracture is a landmark event in the life of men with prostate cancer under ADT. Hip fractures in men over the age of 75 for any cause carry a mortality rate of 30%, while bone fractures in patients with prostate cancer have been associated with adverse overall survival outcomes [27].
- Therefore, androgen ablation is a cause of skeletal-related events associated with significant morbidity and mortality even for patients with nonmetastatic prostate cancer. Risk factors for osteoporotic fracture include the duration of ADT (>3 years), age (mainly through decreased testosterone levels), ethnicity (Caucasian patients are at greater risk), smoking, lower BMI, and medications (e.g., glucocorticoids).
- Lifestyle modifications that apply to all men under ADT irrespective of their bone status include regular light weight lifting or resistance exercises, cessation of smoking, limiting alcohol and caffeine consumption, and vitamin D and calcium supplementation.
- The management of ADT-induced osteopenia and osteoporosis has been a field of significant evolution during the last years, as bone-targeted therapies have recently been the focus of considerable research and drug development. The osteoclast has been recognized as a validated therapeutic target in the management of prostate cancer. Osteoclast inhibition with bisphosphonates reduces the risk for skeletal events in men with castration-resistant prostate cancer metastatic to bone. Osteoclast activity inhibition improves bone mineral density, a surrogate for osteoporotic fracture risk.
- Late generation bisphosphonates such as zoledronic acid seem to markedly reduce bone resorption and increase BMD in prostate cancer patients under ADT. Zoledronic acid has also shown efficacy in preventing bone metastases and skeletal-related events in patients under ADT reducing by seven-fold the risk of pathological fractures after 20 months of treatment. Certain precautions should be taken with the use of zoledronic acid; the risk of osteonecrosis of the jaw is further limited if patients are asked to refrain from dental procedures while on therapy, while the dose and administration of the drug (3 mL intravenous infusion in no less than 15 min) should be adjusted for patients on CKD [28] (details in Chap. 15)
- Denosumab, a human monoclonal antibody that binds to a cytokine ligand responsible for bone tissue replacement is the only FDA-approved medication for preventing ADT-related bone loss [29].

16.3.4 Psychological and Cognitive Effects of Metastatic Prostate Cancer Patients and How to Deal It

- Hormonal therapy can lead to neurologic impairment, including impaired cognitive function, mood, and self-esteem, as well as unfavourable effect like loss of memory depression, anxiety, attention deficit and worse quality of life. Low testosterone levels are strongly linked to depression in elderly men, and testosterone replacement therapy appears to ameliorate depressive symptoms in these patients. As a result, men with prostate cancer appear to be more likely to suffer from depression [30].
- A combination resistance/aerobic training routine can significantly enhance tiredness and cognitive function [31]. Oestrogens can also be used to mitigate the cognitive effects of androgen ablation, although the advantages must be weighed against the well-known danger of cardiovascular events.
- Depression has been recorded after PC diagnosis, although it is unclear if ADT causes or worsens it. Nelson et al. observed no link between ADT treatment and worsening depression symptoms in nondepressed or depressed patients with nonmetastatic prostate cancer. Although antidepressants are rarely effective for ADT-induced depression, they are regularly prescribed to prostate cancer patients [32].

16.3.5 Cardiovascular Complications and Metabolic Syndrome

- Testosterone suppression may produce metabolic changes that contribute to the increased incidence of cardiovascular disease in advanced prostate cancer patients on hormone therapy. Long-term ADT users, particularly older men are more susceptible to the metabolic syndrome. This condition is often associated with an elevated risk of myocardial infarction death, even in the absence of known risk factors like diabetes and pre-existing cardiovascular conditions (Fig. 16.4).
- Hormonal therapy can cause metabolic abnormalities such obesity, insulin resistance, and increased triglycerides, which are akin to the metabolic syndrome. However, it increases subcutaneous fat distribution. Toremifene has been shown to enhance lipid profile, and metformin paired with lifestyle changes is a safe treatment option for undesirable metabolic changes [33]. These drugs are often used in these patients.
- Two studies support the relationship between ADT and cardiovascular morbidity. A recent study validated previous findings that the use of ADT is associated with an earlier onset of fatal acute myocardial infarction in males aged 65 and older who are treated for 6 months. Increased cardiovascular toxicity was assumed to be caused by modifications in lipoproteins, arterial stiffness, and QT interval lengthening [34, 35].

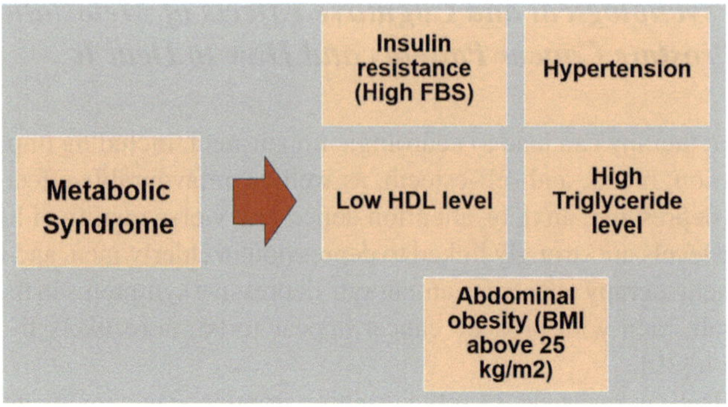

Fig. 16.4 Features of metabolic syndrome. *FBS* fasting blood sugar, *HDL* high-density lipoprotein cholesterol. (Illustration source: Author)

- ADT's catabolic impact can reduce muscle strength. Muscle weakness and poor cognitive function are associated with early loss of functional capacity, limiting independent functioning and lowering quality of life. Daily physical activity is considered the most significant lifestyle change for avoiding these deleterious effects of ADT [33].

16.3.6 Strategies to Curtail the Side Effects of Androgen Deprivation Therapy

- In a study, over 70% of newly treated ADT patients were unaware of potential side effects such as anaemia, cognitive difficulties, hair loss, and depression. Furthermore, more than half were unaware of major potential side effects such as muscle mass loss, osteoporosis, increased fracture risk, weight gain, genital shrinkage, and gynecomastia. A lack of information regarding ADT side effects may be responsible for some of the current significant decreases in patient and partner quality of life. Long-term ADT has been linked to detrimental effects on men's physical as well as psychological well-being. Although there are methods of controlling the vast majority of undesirable effects, compliance is uncertain. There are doubts on whether the patients benefit from this information at all [36].
- Effective solutions for reducing ADT issues have been developed with increased awareness of its negative effects. Improved efforts are to educate patients treatment side effects and coping methods may lead to better outcomes of the ADT. Attempts to alleviate the deleterious effects of ADT include intermittent hormonal therapy and methods of reducing intracellular androgens with maintained circulating testosterone levels [37].
- In some studies it has been found that intermittent ADT reduces early side effects such as hot flashes, sexual activity, and weariness as compared to continuous ADT, but the long-term consequences are unclear [38]. It is fair to expect that

administering LHRH agonists intermittently would alleviate both acute and chronic problems. Intermittent androgen suppression can be used to manage prostate cancer, giving patients a better quality of life throughout therapy while also reducing toxicity and expenses. A recent study found that it also restore potency and resolved anaemia. However, long term studies are required to ensure that intermittent therapy has a deleterious impact on prostate cancer survival.

16.3.7 Gastrointestinal Symptoms

- Anorexia and weight loss are typical in advanced cancer, leading to fatigue, apathy, and family concerns. Megestrol acetate and corticosteroids are commonly used medications to address these symptoms. Megestrol acetate usually works in 1–2 weeks and can boost food intake and improve overall well-being. Corticosteroids, when taken, can have antiemetic and analgesic characteristics, however they are short-lived appetite stimulants with unpleasant medium- to long-term adverse effects.
- Common causes of nausea in advanced prostate cancer include autonomic failure, gastroparesis, constipation, and usage of powerful opioids. To manage symptoms of advanced prostate cancer, it's crucial to address the underlying cause and use an appropriate antiemetic.
- Constipation is common in metastatic ca prostate patients and it can worsen their conditions with symptoms like nausea, urinary retention, lethargy, abdominal pain, and anorexia [39].

16.3.8 Lymphoedema

- Lymphoedema is an accumulation of lymphatic fluids that produces an abnormal swelling in the body. This condition can appear in some people who have been through advanced Prostate Cancer. When lymphatic drainage is blocked, protein-rich fluid gradually builds up and stagnates in the soft tissues where bacterial growth can develop, possibly leading to infection.
- Some people with cancer have a higher risk of developing lymphedema than others. Those who are obese, advancing in age or who have complications after surgery are at a higher risk. In addition to being uncomfortable and easily infected, it can affect relationships with family and friends and cause feelings of isolation and exclusion [40]. Patients with penile and scrotal oedema may experience intense discomfort, which limits their ability to move and makes it difficult for them to urinate while standing. Sexual activity is frequently unfeasible, and the patient's quality of life and self-esteem are frequently negatively impacted by improper perineal cleanliness [41].

- Treatment for lower limb lymphoedema can be challenging and may result from enlarged lymph nodes, pelvic tumours, or obstruction of the inferior vena caval (IVC). Bandaging, manual lymphatic drainage, skin care, early cellulitis treatment, IVC stenting, when necessary, scrotal support, and emotional support are some forms of treatment that may be administered.
- Sadly, medication therapy is mostly ineffectual, and diuretics have minimal anti-edema effects. Deep vein thrombosis (DVT) should always be suspected in conjunction with other conditions, particularly if it is unilateral, uneven, or painful. If this is suspected, ultrasonography Doppler should be used to investigate. In this group of patients, regular low molecular weight heparin is frequently used as treatment rather than warfarin.

16.4 Urinary Obstruction from Advanced Prostate Cancer

- In patients with advanced prostate cancer (PCa), urinary obstruction (UO) can lead to urinary retention requiring indwelling or suprapubic catheterization, which negatively impacts quality of life.
- Palliative options for urinary obstruction (UO) from advanced PCa include surgical and medical treatment. Surgical intervention, including prostatectomy, pelvic exenteration and TURP, may confer undue risk to patients with advanced disease.
- Medical palliation may require months to take effect. Stereotactic Body Radiotherapy (SBRT) is a non-invasive technique that delivers high-dose and highly conformal radiation in ≤5 fractions, often used to treat PCa and metastatic disease.
- SBRT is an effective treatment to palliate UO from locally advanced PCa, leading to removal of urinary catheters and presumed improved in quality of life. Further investigation is required to establish the optimal dose and fractionation of palliative SBRT for UO from PCa [42].

16.5 Music Therapy

- Music therapy has been used successfully for over 30 years as part of palliative care programs for severely ill patients [43–45]. There is nonetheless a lack of high-quality studies that would enable an evidence-based evaluation of its psychological and physiological effects.
- Music therapy is the "systematic use of music within a therapeutic relationship which aims at restoring, maintaining and furthering emotional, physical and mental health" [8]. Music therapy is offered in an extensive range of settings in psychiatric, psychosomatic, neurological, geriatric, paediatric, intensive, and palliative care [9].

- A distinction is drawn between active techniques, in which the patient takes part in producing music using his or her voice or an instrument, and receptive techniques, which involve only listening attentively to music and sounds.
- Music therapy may be an effective nonpharmacological approach to managing distressing symptoms in palliative care patients. The findings also suggest that group music therapy may be a cost-efficient and effective way to support staff caring for palliative care patients. We encourage others to continue developing the evidence base in order to expand our understanding of how music therapy works, with the aim of informing and improving the provision of music therapy for palliative care patients [8].

16.6 The Challenges of the Palliative Care of Advanced Prostate Cancer in Developing Countries

- The following are some of the key challenges in low-income countries:
  - **Limited access to healthcare**: Access to healthcare is a significant challenge in low-income countries. Many patients with advanced prostate cancer in these countries do not have access to appropriate medical care, including palliative care. This is due to a lack of healthcare infrastructure, limited resources, and inadequate healthcare policies [46].
  - **Limited access to pain management**: Pain management is a critical component of palliative care for advanced prostate cancer. However, in low-income developing countries, there is limited access to pain management medications such as opioids. This is due to strict regulations and limited availability of these medications.
  - **Limited availability of trained healthcare professionals**: Palliative care for advanced prostate cancer requires a multidisciplinary team, including physicians, nurses, social workers, and chaplains. However, in low-income countries, there is a shortage of trained healthcare professionals, particularly in rural areas [46].

16.7 Cultural and Religious Beliefs

Patients and their families may refuse palliative care due to cultural or religious beliefs, which may limit their access to appropriate medical care.

- **Stigma and lack of awareness**: Stigma associated with cancer and palliative care can create a barrier to accessing appropriate medical care. Many patients and their families may not be aware of the benefits of palliative care, and may not seek it out due to stigma and lack of awareness.

- **Cost of care**: Many patients and their families cannot afford the cost of medical treatment and palliative care, which limits their access to appropriate medical care.
- **In conclusion**, palliative care for advanced prostate cancer faces several challenges in low-income countries. These challenges include limited access to healthcare, limited access to pain management, limited availability of trained healthcare professionals, cultural and religious beliefs, stigma, lack of awareness, and the cost of care. Addressing these challenges requires a comprehensive approach that involves improving healthcare infrastructure, increasing access to pain management medications, increasing the number of trained healthcare professionals, raising awareness, and developing appropriate healthcare policies.

References

1. Thompson JC, Wood J, Feuer D. Prostate cancer: palliative care and pain relief. Br Med Bull. 2007;83:341–54. https://doi.org/10.1093/bmb/ldm018. Epub 2007 Jul 12.
2. Lec PM, Lenis AT, Brisbane W, Sharma V, Golla V, Gollapudi K, Blumberg J, Drakaki A, Bergman J, Chamie K. Trends in palliative care interventions among patients with advanced bladder, prostate, or kidney cancer: a retrospective cohort study. Urol Oncol. 2020;38(11):854.e1–9. https://doi.org/10.1016/j.urolonc.2020.04.029. Epub 2020 May 17.
3. Cicely Mary Strode Saunders, medical director St Christopher's Hospice 1967–85 (b Barnet, Hertfordshire, 22 June 1918; q St Thomas' Hospital, London, 1957). https://www.bmj.com/content/suppl/2005/07/18/331.7509.DC1.
4. Gleave ME, Bruchovsky N, Moore MJ, Venner P. Prostate cancer: 9. Treatment of advanced disease. CMAJ. 1999;160(2):225–32.
5. Wee B. Oxford textbook of palliative medicine. J R Soc Med. 2004;97(7):356–7.
6. Saunders C. The philosophy of terminal cancer care. Ann Acad Med Singap. 1987;16(1):151–4.
7. Breitbart W. Psychiatric management of cancer pain. Cancer. 1989;63(11 Suppl):2336–42. https://doi.org/10.1002/1097-0142(19890601)63:11<2336::aid-cncr2820631144>3.0.co;2-7.
8. Hole J, Hirsch M, Ball E, Meads C. Music as an aid for postoperative recovery in adults: a systematic review and meta-analysis. Lancet. 2015;386(10004):1659–71. https://doi.org/10.1016/S0140-6736(15)60169-6. Epub 2015 Aug 12. Erratum in: Lancet. 2015 Oct 24;386(10004):1630.
9. Dell'Atti L. Impact of music on anxiety and pain perception among men undergoing prostate biopsy: synthesis of qualitative literature. Complement Ther Clin Pract. 2021;43:101330. https://doi.org/10.1016/j.ctcp.2021.101330. Epub 2021 Feb 6.
10. Lu W, Dean-Clower E, Doherty-Gilman A, Rosenthal DS. The value of acupuncture in cancer care. Hematol Oncol Clin North Am. 2008;22(4):631–48, viii. https://doi.org/10.1016/j.hoc.2008.04.005.
11. White A, Hayhoe S, Hart A, Ernst E, BMAS and AACP (British Medical Acupuncture Society and Acupuncture Association of Chartered Physiotherapists). Survey of adverse events following acupuncture (SAFA): a prospective study of 32,000 consultations. Acupunct Med. 2001;19(2):84–92. https://doi.org/10.1136/aim.19.2.84.
12. MacPherson H, Thomas K, Walters S, Fitter M. A prospective survey of adverse events and treatment reactions following 34,000 consultations with professional acupuncturists. Acupunct Med. 2001;19(2):93–102. https://doi.org/10.1136/aim.19.2.93.
13. Macmillan Cancer Relief. Directory of Complementary Therapy Services in UK Cancer Care: public and voluntary sectors. London: Macmillan Cancer Relief; 2002.

14. Ernst E, Filshie J, Hardy J. Evidence-based complementary medicine for palliative cancer care: does it make sense? Palliat Med. 2003;17(8):704–7. https://doi.org/10.1191/0269216303pm816oa.
15. Filshie J, Hester J. Guidelines for providing acupuncture treatment for cancer patients—a peer-reviewed sample policy document. Acupunct Med. 2006;24(4):172–82. https://doi.org/10.1136/aim.24.4.172.
16. Tazi H, Manunta A, Rodriguez A, Patard JJ, Lobel B, Guillé F. Spinal cord compression in metastatic prostate cancer. Eur Urol. 2003;44(5):527–32. https://doi.org/10.1016/s0302-2838(03)00355-5.
17. Abrahm JL. Assessment and treatment of patients with malignant spinal cord compression. J Support Oncol. 2004;2(5):377–88, 391; discussion 391–3, 398, 401.
18. Nater A, Chuang J, Liu K, Quraishi NA, Pasku D, Wilson JR, Fehlings MG. A personalized medicine approach for the management of spinal metastases with cord compression: development of a novel clinical prediction model for postoperative survival and quality of life. World Neurosurg. 2020;140:654–663.e13. https://doi.org/10.1016/j.wneu.2020.03.098.
19. Tablazon IL, Howard LE, De Hoedt AM, Aronson WJ, Kane CJ, Amling CL, Cooperberg MR, Terris MK, Freedland SJ, Williams SB. Predictors of skeletal-related events and mortality in men with metastatic, castration-resistant prostate cancer: results from the Shared Equal Access Regional Cancer Hospital (SEARCH) database. Cancer. 2019;125(22):4003–10. https://doi.org/10.1002/cncr.32414. Epub 2019 Aug 7.
20. Karling P, Hammar M, Varenhorst E. Prevalence and duration of hot flushes after surgical or medical castration in men with prostatic carcinoma. J Urol. 1994;152(4):1170–3. https://doi.org/10.1016/s0022-5347(17)32530-2.
21. Cervenákov I, Kopecný M, Jancár M, Chovan D, Mal'a M. 'Hot flush', an unpleasant symptom accompanying antiandrogen therapy of prostatic cancer and its treatment by cyproterone acetate. Int Urol Nephrol. 2000;32(1):77–9. https://doi.org/10.1023/a:1007164319067.
22. Harding C, Harris A, Chadwick D. Auricular acupuncture: a novel treatment for vasomotor symptoms associated with luteinizing-hormone releasing hormone agonist treatment for prostate cancer. BJU Int. 2009;103(2):186–90. https://doi.org/10.1111/j.1464-410X.2008.07884.x. Epub 2008 Aug 14.
23. Frisk J. Managing hot flushes in men after prostate cancer—a systematic review. Maturitas. 2010;65(1):15–22. https://doi.org/10.1016/j.maturitas.2009.10.017. Epub 2009 Dec 4.
24. Corona G, Gacci M, Baldi E, Mancina R, Forti G, Maggi M. Androgen deprivation therapy in prostate cancer: focusing on sexual side effects. J Sex Med. 2012;9(3):887–902. https://doi.org/10.1111/j.1743-6109.2011.02590.x. Epub 2012 Jan 16.
25. Corona G, Baldi E, Maggi M. Androgen regulation of prostate cancer: where are we now? J Endocrinol Investig. 2011;34(3):232–43. https://doi.org/10.1007/BF03347072. Epub 2011 Feb 4.
26. Fode M, Sønksen J. Sexual function in elderly men receiving androgen deprivation therapy (ADT). Sex Med Rev. 2014;2(1):36–46. https://doi.org/10.1002/smrj.17. Epub 2015 Oct 19.
27. Gartrell BA, Saad F. Pathologic fracture in patients with metastatic prostate cancer. Curr Opin Urol. 2014;24(6):595–600. https://doi.org/10.1097/MOU.0000000000000105.
28. Saad F, Adachi JD, Brown JP, Canning LA, Gelmon KA, Josse RG, Pritchard KI. Cancer treatment-induced bone loss in breast and prostate cancer. J Clin Oncol. 2008;26(33):5465–76. https://doi.org/10.1200/JCO.2008.18.4184. Epub 2008 Oct 27.
29. Chang H, Marquez J, Chen BK, Kim DM, Cheng ML, Liu EV, Yang H, Zhang L, Sinha M, Cheung A, Kwek SS, Chow ED, Bridge M, Aggarwal RR, Friedlander TW, Small EJ, Anderson M, Fong L. Immune modulation with RANKL blockade through denosumab treatment in patients with cancer. Cancer Immunol Res. 2024;12(4):453–61. https://doi.org/10.1158/2326-6066.CIR-23-0184.
30. Holtfrerich SKC, Knipper S, Purwins J, Castens J, Beyer B, Schlomm T, Diekhof EK. The impact of long-term androgen deprivation therapy on cognitive function and socioeconomic

decision making in prostate cancer patients. Psychooncology. 2020;29(8):1338–46. https://doi.org/10.1002/pon.5442. Epub 2020 Jul 2.

31. McGinty HL, Phillips KM, Jim HS, Cessna JM, Asvat Y, Cases MG, Small BJ, Jacobsen PB. Cognitive functioning in men receiving androgen deprivation therapy for prostate cancer: a systematic review and meta-analysis. Support Care Cancer. 2014;22(8):2271–80. https://doi.org/10.1007/s00520-014-2285-1. Epub 2014 May 25.
32. Nelson CJ, Lee JS, Gamboa MC, Roth AJ. Cognitive effects of hormone therapy in men with prostate cancer: a review. Cancer. 2008;113(5):1097–106. https://doi.org/10.1002/cncr.23658.
33. Conteduca V, Caffo O, Derosa L, Veccia A, Petracci E, Chiuri VE, Santoni M, Santini D, Fratino L, Maines F, Testoni S, De Giorgi U. Metabolic syndrome in castration-resistant prostate cancer patients treated with abiraterone. Prostate. 2015;75(12):1329–38. https://doi.org/10.1002/pros.23014. Epub 2015 May 15.
34. Lai LY, Oerline MK, Caram MEV, Tsao PA, Kaufman SR, Hollenbeck BK, Shahinian VB. Risk of metabolic and cardiovascular adverse events with abiraterone or enzalutamide among men with advanced prostate cancer. J Natl Cancer Inst. 2022;114(8):1127–34. https://doi.org/10.1093/jnci/djac081.
35. Keating NL. Cardiovascular and metabolic diagnoses associated with novel hormonal agents for prostate cancer in nontrial populations. J Natl Cancer Inst. 2022;114(8):1057–8. https://doi.org/10.1093/jnci/djac084.
36. Nguyen PL, Alibhai SM, Basaria S, D'Amico AV, Kantoff PW, Keating NL, Penson DF, Rosario DJ, Tombal B, Smith MR. Adverse effects of androgen deprivation therapy and strategies to mitigate them. Eur Urol. 2015;67(5):825–36. https://doi.org/10.1016/j.eururo.2014.07.010. Epub 2014 Aug 2.
37. Perera M, Roberts MJ, Klotz L, Higano CS, Papa N, Sengupta S, Bolton D, Lawrentschuk N. Intermittent versus continuous androgen deprivation therapy for advanced prostate cancer. Nat Rev Urol. 2020;17(8):469–81. https://doi.org/10.1038/s41585-020-0335-7. Epub 2020 Jun 30.
38. Ahmadi H, Daneshmand S. Androgen deprivation therapy: evidence-based management of side effects. BJU Int. 2013;111(4):543–8. https://doi.org/10.1111/j.1464-410X.2012.11774.x. Epub 2013 Jan 25.
39. Holzbeierlein JM, Castle EP, Thrasher JB. Complications of androgen-deprivation therapy for prostate cancer. Clin Prostate Cancer. 2003;2(3):147–52. https://doi.org/10.3816/cgc.2003.n.022.
40. Huang J, Yu N, Wang X, Long X. Incidence of lower limb lymphedema after vulvar cancer: a systematic review and meta-analysis. Medicine (Baltimore). 2017;96(46):e8722. https://doi.org/10.1097/MD.0000000000008722.
41. Biglia N, Librino A, Ottino MC, Panuccio E, Daniele A, Chahin A. Lower limb lymphedema and neurological complications after lymphadenectomy for gynecological cancer. Int J Gynecol Cancer. 2015;25(3):521–5. https://doi.org/10.1097/IGC.0000000000000341.
42. Boreta L, Raleigh DR, Roach M. Palliation of urinary obstruction from advanced prostate cancer with SBRT. J Clin Oncol. 2017;35(31_Suppl):246. https://doi.org/10.1200/JCO.2017.35.31_suppl.246.
43. Munro S, Mount B. Music therapy in palliative care. Can Med Assoc J. 1978;119(9):1029–34.
44. Bradt J, Dileo C. WITHDRAWN: music therapy for end-of-life care. Cochrane Database Syst Rev. 2014;2014(3):CD007169. https://doi.org/10.1002/14651858.CD007169.pub3.
45. Korczak D, Wastian M, Schneider M. Music therapy in palliative setting. GMS Health Technol Assess. 2013;9:Doc07. https://doi.org/10.3205/hta000113.
46. Bag S, Mohanty S, Deep N, Salins N, Bag S. Palliative and end of life care in India—current scenario and the way forward. J Assoc Physicians India. 2020;68(11):61–5.

Chapter 17
Prevention of Prostate Cancer

Praanjal Gupta and Narmada Gupta

Prevention is better than Cure

17.1 What Is Cancer Prevention?

- Cancer prevention refers to actions performed to reduce the likelihood of developing cancer.

17.2 Why Should We Consider Cancer Prevention?

- Cancer prevention reduces the number of new cases of the disease in a community or group. This should help reduce the number of cancer-related fatalities.

17.3 What Are the Different Levels of Prevention?

- The five stages of a disease's natural history are: underlying, susceptible, subclinical, clinical, and recovery/disability/death.
- To focus on preventing various stages of a disease, corresponding preventive health practices have been organized into analogous stages.
- These four levels of prevention are primordial, primary, secondary, and tertiary.

P. Gupta
Medanta Kidney and Urology Institute, Medanta—The Medicity, Gurugram, Gurugram, India

N. Gupta (✉)
Urology, Medanta Kidney and Urology Institute, Medanta—The Medicity, Gurugram, India

N. Gupta (ed.), *A Clinical Guide for Cancer Prostate Patient*,
https://doi.org/10.1007/978-981-97-5357-4_17

- When combined, these tactics seek to avoid not just the upstream problems of a manifested disease but also the start of disease through risk reduction.
- Quaternary prevention is a newer concept, targeting those having illness without the disease. It aims at alleviating the use of overmedicalization and targets the appropriateness of medical procedures and investigations [1–4].

17.4 What Are Risk Factors and Preventive Factors?

- Scientists study risk factors and preventive variables to stop the development of new malignancies. While many can't be changed, certain cancer risk factors can be avoided.
- For instance, smoking increases the risk of developing specific cancers just as inheriting genes does, but only smoking can be prevented; hence, smoking is a modifiable risk factor while genetics is a non-modifiable risk factor.
- A healthy diet and regular exercise could be preventive measures. Although reducing risk factors and boosting preventive variables may reduce your likelihood of developing cancer, this does not guarantee it [3, 4].

17.5 Risk Factors

Prostate cancer risk may be increased by the following risk factors:

17.5.1 Non Modifiable Risk Factors

17.5.1.1 Age

- Men under the age of 50 are rarely diagnosed with prostate cancer. Men who are older are more likely to develop prostate cancer.

17.5.1.2 Prostate Cancer in the Family/Genetics

- A man has a higher-than-average risk of developing prostate cancer if his father, brother, or son have the disease.
- Compared to a father with a condition, the risk is higher for a brother. If a first-degree relative experienced PCa that manifested early (within 55 years), the risk rises even further.

- Patients with a significant family history of prostate cancer typically present with the disease at a younger age (2.9 years) and with a more locally advanced form of the disease.
- Following a radical prostatectomy, they are also at increased risk of a biochemical recurrence.

17.5.1.3 Ethnicity/Race

- Compared to White men, African American men are more likely to develop prostate cancer and have a higher mortality rate from the condition.
- It is responsible for roughly 37% of all malignancies in male African Americans and 17% of all cancer-related fatalities.
- African Americans have a rate that is 72% greater than that of Whites.

17.5.1.4 Height

- Taller height increases the chances of getting high grade PCa owing to the higher levels of Insulin-like growth factor during puberty.

17.5.2 Modifiable Risk Factors

17.5.2.1 Obesity and Metabolic Syndrome

- BMI >30 was associated with increased chances of getting a high-grade disease.
- Components of metabolic syndrome including hypertension and waist circumference >102 cm enhance the chances of getting prostate cancer.

17.5.2.2 Hormones

- To function properly, the prostate requires male hormones. Testosterone is the primary hormone used in male sex. Male sex traits are created and maintained by testosterone in the body.
- Testosterone is transformed into dihydrotestosterone (DHT) by an enzyme 5alpha-reductase. DHT is necessary for the proper growth of the prostate, but it can potentially enlarge the prostate and contribute to the occurrence of prostate cancer.
- Men who are taking testosterone supplements for testosterone deficiency (hypogonadism) are not at risk of prostate cancer.

17.5.2.3 Sexual Activity

- The risk of prostate cancer is increased by having numerous sexual partners over the course of a lifetime or beginning sexual activity early.
- Prostate cancer risk may be reduced overall by frequent ejaculation, but the prevalence of advanced illness is not increased when ejaculatory frequency is reduced.

17.5.2.4 Infections

- Chlamydia, gonorrhoea, and syphilis infections appear to raise the risk of getting prostate cancer.
- Prostate cancer incidence has been linked to the Human Papilloma Virus (HPV), although the evidence is conflicting.

17.5.2.5 Vitamins and Minerals

- The Selenium and Vitamin E Cancer Prevention Trial (SELECT) found that taking vitamin E by itself increased the risk of prostate cancer. Even when one stops taking vitamin E, there is still a risk.

17.6 Preventive Factors

- The following factors may reduce the risk of prostate cancer and are broadly divided into chemoprevention and dietary factors. Despite the rich evidence, their effect on PCa specific survival and overall survival is still dowdy.

17.6.1 Chemoprevention

17.6.1.1 Finasteride and Dutasteride

- Drugs like finasteride and dutasteride are used to reduce the body's production of male sex hormones. The enzyme that converts testosterone into dihydrotestosterone (DHT) is blocked by these medications. Prostate cancer development may be influenced by DHT levels that are higher than usual.
- It has been demonstrated that using finasteride or dutasteride reduces the risk of developing prostate cancer, but it is unknown whether these medications also reduce the risk of dying from prostate cancer.

17.6.1.2 PCPT Trial and REDUCE Trial (Role in Tertiary Prevention)

- The Prostate Cancer Prevention Trial (PCPT) trial found finasteride to be a potential prostate cancer prevention drug, with lower incidence rates in healthy males aged 55–75 years. However, side effects like erectile dysfunction and larger breasts were more common in finasteride users. Moreover, high-grade prostate cancer was more prevalent in the finasteride group, but after 18 years of follow-up, there was no discernible difference in the rates of overall survival or survival following a prostate cancer diagnosis.
- The Reduction by Dutasteride of Prostate Cancer Events (REDUCE) trial found fewer prostate cancers in men taking dutasteride, with fewer aggressive prostate cancers but more side effects.

17.6.1.3 Aspirin, NSAIDs

- The risk of PCa was considerably decreased with regular aspirin use. Similarly, prolonged use of nonsteroidal anti-inflammatories was discovered to be protective against PCa in a meta-analysis of 43 observational studies.

The following have been proven not to affect the risk of prostate cancer, or their effects on prostate cancer risk are not known:

17.6.1.4 Selenium and Vitamin E

- The Selenium and Vitamin E Cancer Prevention Trial (SELECT) trial examined the potential of vitamin E and selenium in preventing prostate cancer in healthy men aged 55–65. Results showed selenium alone or combined with vitamin E did not reduce prostate cancer risk.

17.6.1.5 Folate

- Folate is a type of vitamin B found naturally in meals, including green vegetables, beans, and orange juice.
- Folic acid is a synthetic version of folate found in vitamins and fortified meals such as whole-grain breads and cereals.
- A long-term (10-year) study found that males who ate enough folate had a decreased risk of prostate cancer. Men who used 1 milligram (mg) folic acid supplements, on the other hand, had an elevated risk of prostate cancer.

17.6.1.6 Statins

- A recent meta-analysis of reviews revealed a decreased incidence of PCa with longer statin treatment and greater doses. Comparatively, some investigations have discovered a weak or non-existent correlation between statin use and the total incidence of PCa [4–7].

17.6.1.7 Metformin

- Intake of metformin is associated with reduced risk of prostate cancer and a further reduction in prostate cancer mortality. Studies have found further role of metformin in improving metabolic syndrome and as an adjuvant agent with hormonal treatment for prostate cancer.

17.6.2 Diet: What or What Not to Eat?

- It is not known if decreasing fat or increasing fruits and vegetables in the diet helps decrease the risk of prostate cancer or death from prostate cancer.
- The associations of dietary factors with prostate cancer are more conjectural; however, the below-mentioned points can be considered.

17.6.2.1 Say Yes to Them

- Mediterranean diet—Nuts, fruits, vegetables, legumes, red wine, fish, and olive oil are all abundant in the Mediterranean diet, which is regarded as being beneficial.
- Meals rich in omega-3 fatty acids, phenolic compounds, and oleic acid are just a few examples of anti-oxidant and anti-inflammatory chemicals.
- Higher retinol-containing foods are protective.
- Cruciferous vegetables—especially broccoli, which is a significant source of sulforaphane. Also, cruciferous vegetables are a rich source of isothiocyanates which is considered protective.
- Coffee consumption may reduce the risk of getting prostate cancer.
- Lycopene-rich diets may lower prostate cancer risk, but research on their potential benefits has not been validated. Consuming lycopene supplements has not been proven to lower the risk.
- Phytoestrogens have been linked to a lower risk of PCa. Foods including soybeans, lentils, tofu, peanuts, chickpeas, and kidney beans contain phytoestrogens.

17.6.2.2 Say No to Them

- Saturated animal fats—Increased levels of circulating androgens have been demonstrated to enhance PCa cell proliferation in response to saturated animal fats. Red meats, total meats and processed meats all increase the risk of getting prostate cancer.
- Trans fatty acids—Consumption of trans fatty acids, which are frequently used in the manufacture of ultraprocessed foods, was linked to an increased risk of PCa.
- Multivitamins—Regular multivitamins intake doesn't increase prostate cancer risk; a study found men taking them more than seven times per week had higher risk.
- Vitamin D—Both low and high levels of vitamin D portend a risk of prostate cancer.
- Prostate cancer patients with vitamin D deficiencies have a higher overall and cancer-specific mortality rate. This suggests that vitamin D supplements may be helpful for prostate cancer patients who are deficient in the vitamin.
- Calcium and dairy—A diet heavy in dairy and calcium may increase the risk of prostate cancer by a tiny amount [5–9].

17.6.2.3 Strategy to Prevent Prostate Cancer: A Primer on Lifestyle Interventions

- It is scientifically proven that healthy diet and lifestyle practices can aid in slowing down the growth and progression of prostate cancer.
- A healthy lifestyle can also help manage many of the treatment's side effects as well as lower the risk of the cancer growing or coming back.

17.6.2.4 Maintain a Healthy Body Weight

- The risk of prostate cancer recurrence increases in men who are overweight or obese and also can decrease the survival rate.
- The risk of spread and recurrence is lower among person with healthy weight and it also improves survival.
- BMI = 18.5–24.9—Healthy Body Weight.
- BMI >25.0, Waist circumference is >102 cm—Weight Loss is Recommended.
- Hence, one should prevent weight gain, and achieve a healthy weight with a balanced diet along with regular physical activity.

17.6.2.5 Dietary Changes

- Changes in daily eating habits help decrease the risk of cancer coming back or getting worse and also support recovery from prostate cancer.

- Some of the dietary recommendations that lower the risk of prostate cancer progression are:
  - Includes fruit, vegetables, and protein-rich foods in the diet.
  - Cut down on unhealthy foods and drinks (sugary drinks).
  - Ground flaxseed can lower the risk of cancer progression.
  - Foods rich in selenium, such as fish, cottage cheese, or a selenium supplement, may be beneficial.
  - Foods containing B-carotene like carrots, spinach, sweet potatoes, and broccoli are encouraged.
  - Maintain adequate Vitamin D levels for healthy bones.
  - Limit consumption of red meats, processed meats, and salty foods.
  - Choose fish, lean poultry, or plant-based proteins.

17.6.2.6 Regular Physical Activity

- Exercise is beneficial for men with prostate cancer. Regular physical activity offers benefits like:
  - Increase overall wellbeing and quality of life
  - Help minimize the side effects of prostate cancer treatments.
  - Improve bone health, prevent or minimize the loss of muscle mass
  - Help maintain a healthy body weight.
  - Pelvic floor muscle exercise helps control urination and empty bowels. It is recommended before and after radiotherapy or surgery.

17.6.2.7 Yoga

- In males with prostate cancer, preoperative yoga exercise increases quality of life, boost the immune system, and reduce inflammation. It also improves overall health and wellbeing.

17.6.2.8 Reduce Alcohol Intake

- Drinking too much alcohol can make you gain weight and cause other health problems.

17.6.2.9 Quit Smoking

- Smoking makes prostate cancer more likely to grow, spread, and recur.
- Quitting smoking decreases the risk and helps with the side effects of treatment.

17.6.3 *Living with Prostate Cancer: A Portrait from the Patient Aspect (Tertiary Prevention)*

Prostate cancer can affect one's life both physically, emotionally, and practically, as well as the lives of family members.

Some of the common thoughts and feelings are:

- Shock, fear or anger
- Worries about side effects
- Sense of loss
- Denial
- Feeling alone
- Frustration and disappointment
- Mood swings
- Stress
- Anxiety

It is normal to feel all these emotions. It might stay with some of them, while some men find they gradually change with time.

17.6.4 *Relationships and Family Life*

- Prostate cancer can affect life, relationships, friendships, and roles within the family.
- One's plans get interrupted or priorities change after the diagnosis of prostate cancer.
- It can also bring some couples and families closer together.
- Prostate cancer and its treatments can also affect relationships and sexual life one might need support for relationship and sexual issues.

Patients are at risk of developing

- Depression
- Erectile dysfunction
- Infertility
- Loss of libido
- Anxiety

Some of the ways which can help one living with prostate cancer

- Talking about things with loved one or a health professional
- Getting help with practical matters such as work, money or household tasks
- Finding ways to manage or treat side effects
- Learning more about prostate cancer
- Getting all the support one need as a family

- Developing a wider support network including other family, friends or health professionals

17.6.5 *Advice for Caregivers: They May Be Crucial in Tertiary Prevention*

A caregiver is essential in providing physical and emotional support.

Some of the tips to get started for the caregiver role:

- Go to doctor appointments.
- Take notes. As, people coping with cancer may not hear everything and miss important information.
- If asked, help discover treatment options.
- Take time to do what nurtures you.
- Take a walk, talk to friends, read, and play music.
- If loved one is overwhelmed, talk to him and his doctor about counselling.
- All prostate cancer treatments have side effects; it is important to discuss with the loved one how you'll be involved in supporting.
- Remember, both patients & caregivers will experience a wide range of emotions.
- Do not suppress these feelings, talk about them together and know that they are normal [4, 6, 8–11].

17.6.6 *Quaternary Prevention*

- This occurs when the disease is absent but the illness is present.
- This revolves around the concept of narrative based medicine, which leads to the prevention of unnecessary medical interventions which is in concordance with the patient.
- Efficient communication and trust of the patients are the guiding lights for quaternary prevention.
- An illustration of quaternary prevention is the recommendation of preventive measures that have been shown to be effective. In terms of diagnosis, quaternary prevention includes things like avoiding unfounded screenings for prostate cancer.

17.7 Conclusion

- Prevention of prostate cancer shall be an intricate and organised blending of multiple disciplines including urology, primary care, epidemiology, and health policy.
- PCa does appear to be less common when using dietary and pharmacological supplements. Their effect on general survival and PCa-specific survival is still unknown.
- Because of the potential for more advanced disease at diagnosis and the associated toxicities, use of chemopreventive methods has not been widely accepted.
- The use of protective dietary agents should be encouraged as they would upend the overall health of older men, apart from their beneficial effect against PCa.
- Healthy active life style keeps morals high to fight the cancer
- New techniques to prevent prostate cancer are being explored in clinical trials, which are used to study cancer prevention strategies.

References

1. Culp MB, Soerjomataram I, Efstathiou JA, Bray F, Jemal A. Recent global patterns in prostate cancer incidence and mortality rates. Eur Urol. 2020;77(1):38–52. https://doi.org/10.1016/j.eururo.2019.08.005.
2. Kisling LA, Das JM. Prevention strategies [updated 2023 May 1]. In: StatPearls. Treasure Island, FL: StatPearls Publishing; 2023. Available from: https://www.ncbi.nlm.nih.gov/books/NBK537222.
3. AbdulRaheem Y. Unveiling the significance and challenges of integrating prevention levels in healthcare practice. J Prim Care Community Health. 2023;14:21501319231186500. https://doi.org/10.1177/21501319231186500.
4. Prostate Cancer Prevention (PDQ®). NCI. 2023 [cited 2023 Jul 23]. Available from: https://www.cancer.gov/types/prostate/hp/prostate-prevention-pdq.
5. Bergengren O, Pekala KR, Matsoukas K, et al. 2022 Update on prostate cancer epidemiology and risk factors—a systematic review. Eur Urol. 2023;84(2):191–206. https://doi.org/10.1016/j.eururo.2023.04.021.
6. Schaeffer EM, Srinivas S, Adra N, et al. NCCN Guidelines® insights: prostate cancer, version 1.2023. J Natl Compr Cancer Netw. 2022;20(12):1288–98. https://doi.org/10.6004/jnccn.2022.0063.
7. EAU guidelines on prostate cancer. Uroweb. European Association of Urology [cited 2023 Jul 23]. Available from: https://uroweb.org/guidelines/prostate-cancer.
8. Ornish D, Weidner G, Fair WR, et al. Intensive lifestyle changes may affect the progression of prostate cancer. J Urol. 2005;174(3):1065–70. https://doi.org/10.1097/01.ju.0000169487.49018.73.
9. Wilson KM, Giovannucci EL, Mucci LA. Lifestyle and dietary factors in the prevention of lethal prostate cancer. Asian J Androl. 2012;14(3):365–74. https://doi.org/10.1038/aja.2011.142.
10. Living as a prostate cancer survivor—after prostate cancer treatment [cited 2023 Jul 23]. Available from: https://www.cancer.org/cancer/types/prostate-cancer/after-treatment/follow-up.html.
11. Prostate cancer: a caregiver's guide [cited 2023 Jul 23]. Available from: https://www.urology.org/caregiverguide/.

Chapter 18
Prostatic Biopsy

Somendra Bansal and Narmada Gupta

Prostatic Biopsy is the corner stone for the diagnosis of Cancer Prostate (NPG)

- Prostate cancer is the commonest cancer (27%) in male in the United States and it accounts second most common cause of cancer related mortality (10%) [1]. It is also one of the five leading cancers among males in the most of registries of India. Prostate cancer should be suspected in case of abnormal digital rectal examination (DRE) and increased prostate specific antigen (PSA) (>4 ng/mL). Prostate biopsy is needed for confirmation of cancer.
- There is a myth that Prostatic biopsy may spread cancer in the body. This is wrong. In the past, usually prostate cancer patients presented late and also with metastatic disease. After biopsy, when they were investigated further, these metastatic sites were detected. This gives an impression that prostatic biopsy can spread the tumour.

18.1 Types of Biopsy

- Initially finger guided biopsy were done but with the availability of Ultrasound machines, ultrasound guided biopsy is a routine. End-fire and side-fire ultrasound probes along with transrectal and trans perineal approaches are used for prostate biopsy.

S. Bansal
Urology and Renal Transplant, SMS Medical College and Hospitals, Jaipur, India

N. Gupta (✉)
Urology, Medanta Kidney and Urology Institute, Medanta—The Medicity, Gurugram, India

N. Gupta (ed.), *A Clinical Guide for Cancer Prostate Patient*,
https://doi.org/10.1007/978-981-97-5357-4_18

- Transrectal ultrasound (TRUS) prostate biopsy is commonly used for diagnosis of prostate cancer, because of its easily availability and real time in nature. But it has some limitations like low diagnostic yield and sampling error. It has tendency to miss the lesions located anteriorly and at apex of the prostate. It has risk of infection as needle passes through the rectum.
- At present, use of 10 to 12-core systematic biopsy is recommended for initial prostate biopsy incorporating apical and far-lateral samples [2, 3]. TRUS guided prostate biopsy had 39–52% sensitivity and approximately 80% specificity for detection of prostate cancer [4].
- Repeat prostate biopsies were often required because of high false-negative rates of TRUS guided biopsies. Up to 40% of prostate cancer detected by TRUS guided biopsy may upgrade in higher Gleason score on final surgical histopathology. Thus, it is possible that the most clinically significant prostate cancer may not accurately stratified by this method [5, 6]

18.2 Trans Perineal Prostate Biopsy (TPP)

- TPP biopsy was first described in 1970s, but recently it is more widely accepted method of prostate biopsy due to high sensitivity in detecting anterior lesions of prostate and lower rate of sepsis as compared to TRUS biopsies.

Trans perineal prostate biopsy is specifically useful in following patients:

- History of sepsis and rectal bleeding after previous transrectal prostate biopsy
- Prostatitis and inflammatory bowel disease
- Previous negative transrectal biopsy with suspicious of lesion in apex and anterior prostate
  - It has potential for reduced infections and other complications and improved identification of apical tumours
  - Following absent rectum due to previous surgery/congenital

    In trans perineal biopsy, ultrasound probe is placed in rectum which guide the biopsy needle that passes through perineal skin into the prostate

Recently multiparametric (mp) magnetic resonance imaging (MRI) has emerged as imaging method of choice for diagnosis of prostate cancer. It is used for detection, localization and biopsy of suspicious prostate lesions [7].

MRI suspicious lesions can be biopsies by three techniques:

- MRI cognitive biopsy (MRI suggested)
- MRI biopsy in gantry
- MRI-TRUS fusion biopsy
- **Real time MRI biopsy in gantry** is cumbersome and time consuming as whole procedure is done in MRI suite. It is not popularly done [8, 9].

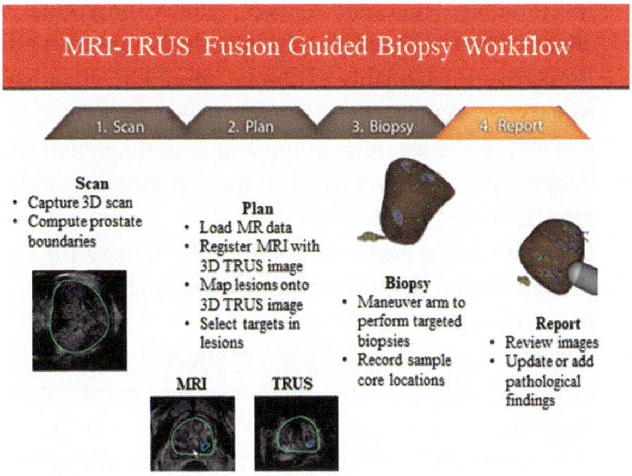

Fig. 18.1 MRI-TRUS fusion guided biopsy workflow

- **MRI suggested cognitive guidance** can be performed and lesions can be targeted accordingly. In absence of fusion plate form, this type of the biopsy is done.
- In MRI-TRUS fusion prostate biopsy, previously acquired MRI images are fused with real time TRUS by a semi-robotic targeted biopsy software. This technique combines the benefit of both high resolution of MRI and real time ultrasound. This fusion biopsy accurately target MRI suspicious lesion and decrease the sampling errors. Fusion biopsy has role in patients with persistently raised serum PSA and previous negative TRUS prostate biopsy (Fig. 18.1) [10–12].

18.3 Percutaneous Trans Gluteal Prostate Biopsy

- CT guided biopsy/PSMA PET/CT-guided trans gluteal is another approach for prostatic biopsy in special situation of absent rectum or previously negative biopsy.
- It is feasible and safe technique with high diagnostic yield. It also has lower morbidity due to bacterial infection [13].

18.4 Common Steps for Prostatic Biopsy Procedure

- Patients should have normal coagulation profile and sterile urine culture before undergoing biopsy. Patients taking anticoagulant should be stopped at least 5–7 days prior to biopsy. Prostate biopsy should not be done in acute urinary infection and coagulopathy.
- Counselling of the patient is done and procedure is explained. Consent is taken for the procedure. Usually, procedure can be done as day care or admission in high

risk patients with comorbidities or unwilling under local anaesthesia. Proctoclys enema and antibiotic prophylaxis are given to patients before prostate biopsy. Prophylactic injectable antibiotic should be given within 1 h before the biopsy

- Mostly, procedure is done under local anesthesia, however if patient is not comfortable, it can be done in done in general, spinal anesthesia.
- Patients are usually placed in the left lateral decubitus position with knees and hips flexed 90° (Fig. 18.2).
- DRE should be performed before the biopsy for clinical evaluation of prostate. Perineum and rectum inside and outside cleaned with betadine solution (10% povidone-iodine)
- Intrarectal anesthetic jelly 2% is injected in the rectum and wait for 5–10 min for its action. Under TRUS guidance, peri prostatic block is also given using 2% lidocaine.
- Ultrasound probe is inserted in the rectum after proper lubrication and prostate is scanned for identification of landmarks and suspicious lesions.
- Prostate specific landmarks are urethra, transitional and peripheral zone and verumontanum. Seminal vesicles have moustache appearance at base of prostate.
- Patient specific landmarks are calcification, cysts and hypoechoic lesions. Prostate volume (mL) is calculated using formula height × width × length × 0.52.
- Under the ultrasound guidance systemic 12 core biopsy is performed. Prostate is divided into right and left lobes and each lobe is further divided into six regions: base, mid and apex (both lateral and medial).
- Targeted biopsy from MRI suspicious lesions should be performed prior to systemic 12 core biopsy to avoid impairment of target area. The number of cores taken should balance the detection of clinically significant prostate cancer whilst minimizing side effects associated with increased sampling numbers.
- Cores from each site are sent separately for histopathological examination in 10% formalin. Care must be taken to labelling the sample accurately from the prostate area they are taken. The procedure usually takes 10–15 min.
- In MRI-TRUS fusion biopsy, samples are taken from suspicious lesions which are marked and scored in MRI. Lesions located in MRI are fused onto real-time

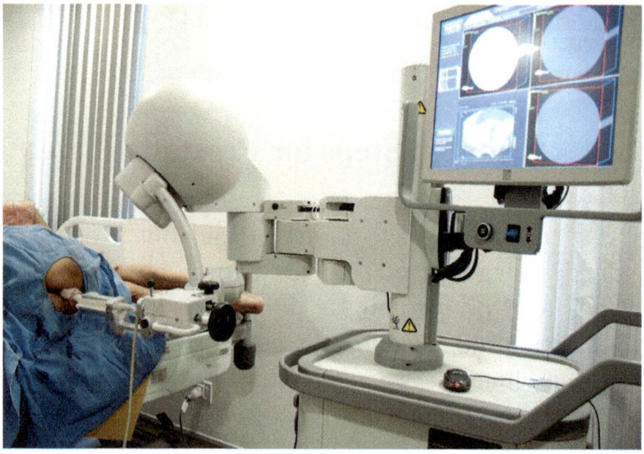

Fig. 18.2 Patient undergoing MRI-TRUS fusion prostate biopsy in left lateral position. (Source: Personal collection)

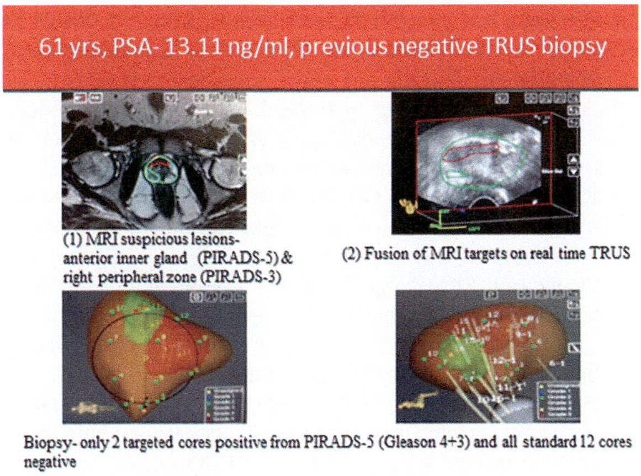

Fig. 18.3 Lesions located in MRI are fused onto real-time TRUS imaging by advanced 3D semi-robotic targeted prostate biopsy software. (Source: Personal collection)

TRUS imaging by advanced 3D semi-robotic targeted prostate biopsy software. The fusion result in creation of three-dimensional reconstruction of prostate. The operator is guided to these specific locations on TRUS imaging by tracking the position of the biopsy needle with mechanical arm (Fig. 18.3).

- Prostate biopsy is the day care procedure. Patient is asked to void before leaving the hospital. Patient may need catheterization if he fails to void. Patient is asked to seek medical attention immediately if he develops complication. Non-steroid anti-inflammatory drugs are prescribed for pain management.

18.5 Complications of Prostate Biopsy

- After prostate biopsy, bleeding is the most common complication, which is usually minor and self-limiting. Bleeding can occur in form of haematuria (10–84%), haematospermia (1.3–45%) or rectal bleeding (1.1–93%). Significant haematuria requiring hospitalization occurs in less than 1% of patients. Incidence of infection requiring hospitalization is 0–6.3% [14]. Sometimes, patient may need hospitalization. Patient should be informed for these complications and management.

References

1. Siegel R, Ma J, Zou Z, Jemal A. Cancer statistics, 2014. CA Cancer J Clin. 2014;64(1):9–29.
2. Bjurlin MA, Carter HB, Schellhammer P, Cookson MS, Gomella LG, Troyer D, et al. Optimization of initial prostate biopsy in clinical practice: sampling, labeling and specimen processing. J Urol. 2013;189(6):2039–46.
3. Bjurlin MA, Wysock JS, Taneja SS. Optimization of prostate biopsy: Review of technique and complications. Urol Clin North Am. 2014;41(2):299–313.

4. Pondman KM, Futterer JJ, ten Haken B, Kool LJS, Witjes JA, Hambrock T, et al. MR-guided biopsy of the prostate: an overview of techniques and a systematic review. Eur Urol. 2008;54(3):517–27.
5. Isariyawongse BK, Sun L, Banez LL, Robertson C, Polascik TJ, Maloney K, et al. Significant discrepancies between diagnostic and pathologic Gleason sums in prostate cancer: the predictive role of age and prostate-specific antigen. Urology. 2008;72(4):882–6.
6. King CR, McNeal JE, Gill H, Presti JC Jr. Extended prostate biopsy scheme improves reliability of Gleason grading: implications for radiotherapy patients. Int J Radiat Oncol Biol Phys. 2004;59(2):386–91.
7. Stephenson SK, Chang EK, Marks LS. Screening and detection advances in magnetic resonance image-guided prostate biopsy. Urol Clin North Am. 2014;41(2):315–26.
8. Yacoub JH, Verma S, Moulton JS, Eggener S, Oto A. Imaging-guided prostate biopsy: conventional and emerging techniques. Radiographics. 2012;32:819–37.
9. Hambrock T, Somford DM, Hoeks C, Bouwense SA, Huisman H, Yakar D, et al. Magnetic resonance imaging guided prostate biopsy in men with repeat negative biopsies and increased prostate specific antigen. J Urol. 2010;183(2):520–7.
10. Natarajan S, Marks LS, Margolis D, Huang J, Macairanb ML, Lieub P, et al. Clinical application of a 3D ultrasound-guided prostate biopsy system: biopsy tracking and lesion targeting via real-time MRI/ultrasound fusion. Urol Oncol. 2011;29(3):334–42.
11. Pinto PA, Chung PH, Rastinehad AR, Baccala AA, Kruecker J, Benjamin CJ, et al. Magnetic resonance imaging/ultrasound fusion guided prostate biopsy improves cancer detection following transrectal ultrasound biopsy and correlates with multiparametric magnetic resonance imaging. J Urol. 2011;186(4):1281–5.
12. Sonn GA, Natarajan S, Margolis DJ, Macairan M, Lieu P, Huang J, et al. Targeted biopsy in the detection of prostate cancer using an office-based MR-US fusion device. J Urol. 2013;189(1):86–91.
13. Kumar R, Mittal B, Singh SK, Singh H, Bhattacharya A. Percutaneous transgluteal Ga-68 PSMA PET/CT guided prostate biopsy: is it the future? J Nucl Med. 2022;63(Suppl 2):2542.
14. Loeb S, Vellekoop A, Ahmed HU, Catto J, Emberton M, Nam R, et al. Systematic review of complications of prostate biopsy. Eur Urol. 2013;64(6):876–92.

Chapter 19
Radical Prostatectomy Surgery

Mayank Agrawal, Gagan Gautam, and Narmada Gupta

Robotic Radical Prostatectomy is the standard of surgical Care in management for localized carcinoma Prostate (NPG)

19.1 What Is Radical Prostatectomy Surgery?

- A radical prostatectomy is a surgical procedure whereby the prostate gland is removed. In this surgery the prostate gland, seminal vesicles, surrounding tissues, and, on most occasions, regional lymph nodes are removed to provide the best possible chance of removing all the cancer. The seminal vesicles are glandular structures lying behind the prostate which may be invaded by prostate cancer (Fig. 19.1). The procedure to remove the lymph nodes is called "pelvic lymph node dissection [1]."
- The aim of radical prostatectomy is to completely eradicate the disease from the body which can be possible if disease is localized to the prostate gland (Figs. 19.2 and 19.3).
- Radical prostatectomy is a complex surgical procedure which needs a balance between removal of disease from the body but to preserve the sexual nerves and sphincteric structures to achieve a good oncological and functional outcomes like continence and sexual activities.

M. Agrawal
Vattikuti Foundation, Southfield, MI, USA

Uro-Oncology and Robotic Surgery, Medanta, Gurugram, Gurugram, India

G. Gautam (✉)
Uro-Oncology and Robotic Surgery, Medanta, Gurugram, Gurugram, India
https://www.gagangautam.com

N. Gupta
Urology, Medanta Kidney and Urology Institute, Medanta—The Medicity, Gurugram, India

N. Gupta (ed.), *A Clinical Guide for Cancer Prostate Patient*,
https://doi.org/10.1007/978-981-97-5357-4_19

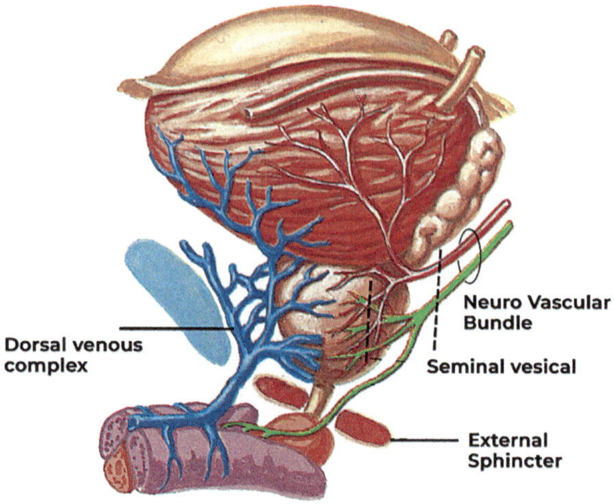

Fig. 19.1 Anatomy of prostate

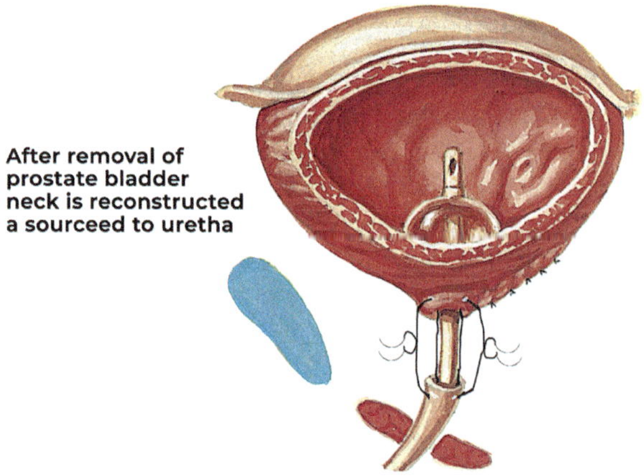

Fig. 19.2 After radical prostatectomy. (Source: NPG)

- There are several approaches for doing a radical prostatectomy. Open, Laparoscopic, and Robot-assisted radical prostatectomy. Radical prostatectomy can be done through the abdomen or perineum.
- In open surgery, the surgeon cuts the abdominal wall (transabdominal) or the perineum (perineal) to access the prostate directly. Patrick Walsh from USA described open transabdominal Anatomical radical prostatectomy in 1982. Since then, it has become a popular operation. This open approach has increased blood loss and morbidity [2, 3].
- In laparoscopic surgery, through small holes in the abdomen, the camera and instruments are inserted to remove the prostate gland. Ralph Clayman in 1992 described Laparoscopic approach which was popularized by a group of French surgeons. It has advantages of minimally invasive surgery but technically difficult to perform and take longer period. It is limited to few expert laparoscopic surgeons all over the world [2].

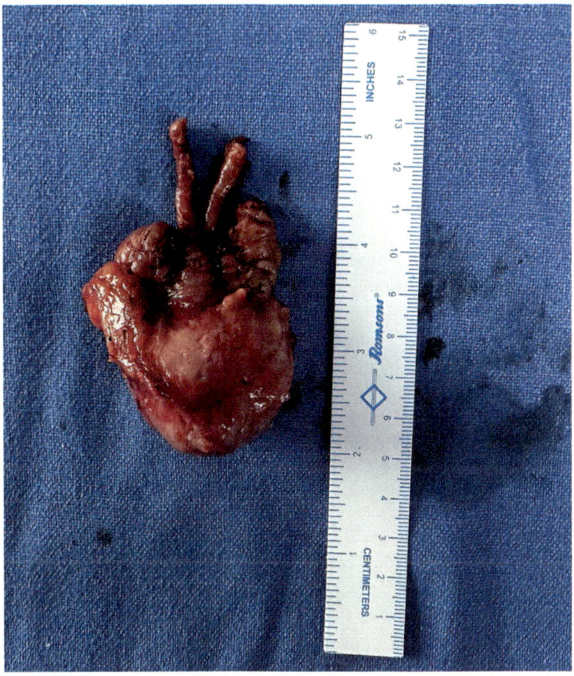

Fig. 19.3 Radical prostatectomy specimen. (Source: NPG)

- Robotic radical prostatectomy started in 2000 but popularized by Dr. Mani Menon from USA in 2002. Soon, it became popular all over the world and today more than 90% radical prostatectomy are done robotically. Several urologist modified and refined various steps of the surgery to improve functional and oncological outcomes. Many urologist from all over the world adopted this technology because of less steep learning curve and convenience of doing this surgery with all advantages of minimally surgery and less blood loss and early recovery and return to work. They have also changed their approach from open, laparoscopic to robotics.
- Robotic surgery involves the use of the robotic surgical unit in performing this complex surgical procedure through small holes. Originally developed for space and military uses, this advanced technology allows the operating surgeon to sit behind a console and control the robot that is next to the patient to perform surgery.
- The patient cart of the robotic system has four robotic arms; three for instruments and one for a 3-D camera with high magnification allowing the surgeon to see inside the abdomen. The instruments used are approximately 7 mm in width. The instruments have a greater range of movement compared to the human hand to be able to carry out more controlled and precise movements.
- At all times during the procedure the instruments are controlled by the surgeon (who does the operation) and the robot cannot perform any step of the surgery on its own. Surgeon's skill and experience is important for good outcome (Figs. 19.4, and 19.5).

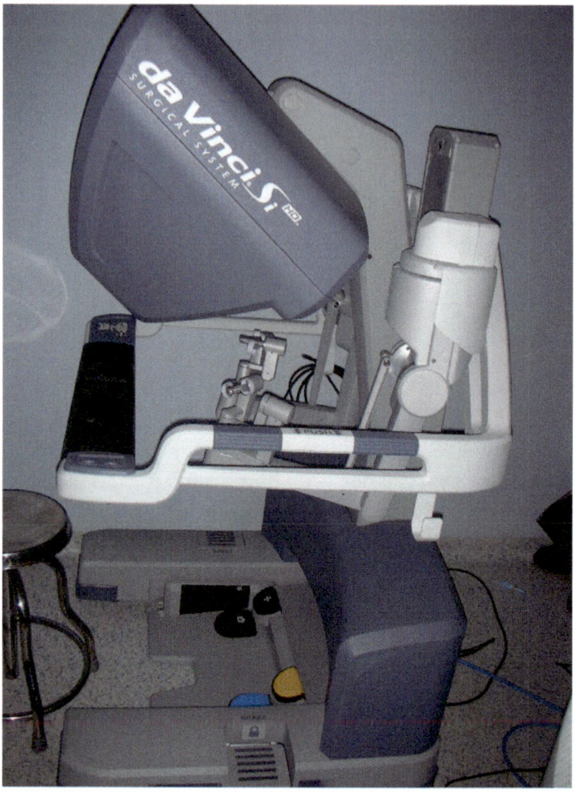

Fig. 19.4 Robotic system. (Courtesy: Medanta, The Medicity)

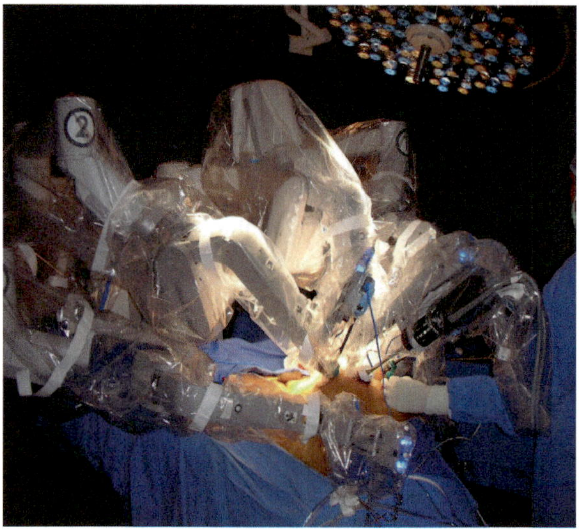

Fig. 19.5 After docking of the robot. (Courtesy: Medanta, The Medicity)

19.2 What Are the Advantages of the Robotic Approach over the Laparoscopic or Open Approach?

- The robotic approach is associated with less blood loss, a shorter hospital stay, and faster recuperation.
- Preservation of nerves, urine control mechanism and other vital structures around the prostate may be better via a robotic approach because of higher magnification, better control of instruments, and higher precision.

19.3 Preparation Before Surgery?

- During OPD evaluation basic blood tests and radiological tests (MRI/CT/PSMA PET Scans) are performed to determine the extent of the disease.
- Patient is evaluated by the anaesthesiologist, cardiologist, and other specialists (depending upon the other medical issues) to assess the patient's fitness for surgery and anaesthesia.
- Patient is admitted to the hospital one day prior to or on the day of the surgery.
- Eight hours of fasting is needed prior to the procedure. Bowel preparation in the form of a rectal enema to cleanse the bowel may be given a few hours prior to the surgery.
- Injection of low molecular weight heparin (anti-coagulant) may be given 12 h prior to the surgery to avoid the risk of deep venous thrombosis. This is a routine protocol that is followed prior to other long-duration surgeries, pelvic, and other cancer surgeries.

19.4 What Happens During the Procedure?

- The surgery is performed under full general anesthesia. The total surgery and anesthesia duration is usually around 4 h but may vary.
- Single shot of intravenous antibiotics is given at the time of induction (30 min prior to the start of surgery).
- The Robot-assisted Radical Prostatectomy is commonly performed using the Da Vinci® system.
- After the complete prostate gland along with surrounding tissues is removed, the bladder and the urethra are attached together (Fig. 19.2).
- A urine catheter is kept for the healing of the urethra and the bladder. The catheter is removed after 7–10 days. Also, a drain tube (the tube that drains fluid/blood accumulations) is left in place for 1 or 2 days.

19.5 What Are the Complications and Quality of Life Issues with This Surgery? [4–6]

19.5.1 Common Side-Effects (Greater Than 1 in 10)

- Surgical site pain—managed with intravenous/oral medications.
- Discoloration of skin at the surgical site (Ecchymosis)—resolves spontaneously.
- Temporary difficulties with urinary control (usually recovers in 6–12 months).
- Impairment of penile erections.
- Inability to ejaculate (occurs in 100% of patients).
- Paralytic ileus—commonly resolves spontaneously in 24–48 h.
- Penile or scrotal skin swelling—resolves spontaneously.

19.5.2 Occasional Side-Effects (Between 1 in 10 and 1 in 50)

- Bladder neck narrowing (Vesico-urethral stenosis) resulting in poor urinary stream and requiring further surgery (2–5%).
- Long-term urinary incontinence (temporary or permanent) requiring pads or further surgery (2–5%).
- Fluid (Lymph) collection in the pelvis if lymph node dissection is performed.
- Development of a hernia at the incision site after the operation.
- Injury to obturator nerve

19.5.3 Rare Side-Effects (Less Than 1 in 50)

- Anesthetic or cardiovascular problems possibly requiring intensive care admission (including chest infection, pulmonary embolus, stroke, deep vein thrombosis, heart attack, and death).
- Blood loss requiring transfusion or repeat surgery.
- Rectal injury requiring a temporary colostomy (less than 1 in 300).

19.6 Care After the Surgery?

- Once the surgery is complete, the patient is taken to the post-operative recovery area and subsequently to the room after 2 h.
- Patients are started on oral fluids shortly after the procedure and will be allowed to eat solid food thereafter.

- Patients are mobilized out of bed and allowed to walk on the day of the surgery with assistance.
- Since the surgery is performed through small incisions, most patients experience much less pain than with open surgery.
- After surgery patients receive intra-venous Acetaminophen (anti-inflammatory and analgesic medication) which is sufficient in the majority of the patients for pain control. For some men, non-steroidal anti-inflammatory drugs in the form of Diclofenac (patch/intravenous) are added if the pain is more severe.
- Abdominal drain is usually removed the next day and the site is covered with a small bandage.
- All other abdominal incisions are closed with absorbable sutures, so no sutures or clips need to be removed. These usually fall off by themselves in 4–6 weeks.
- The incisions are covered with bandages/Dermabond (bio glue). The glue will gradually wear away, within a week or 2.
- Most patients are fit to go home the next day after the surgery. Some will stay a second day, especially if the surgery finished later in the day or if there is post-operative paralytic ileus, which usually resolves spontaneously in 24–48 h.

19.7 What to Expect After Going Home?

- Patients are discharged home with a supply of pain-killing medication, as well as some laxative medication to keep their bowel moments regular. The pain usually disappears after 1 week.
- Physical activity in the form of light walking, stretching, or climbing a staircase is allowed after the procedure. The intensity of the physical activity can be increased gradually after 4 weeks from surgery. Any activity that causes pain should be avoided.
- Patients are allowed to take a shower or bath after the surgery as the stitches are dissolvable.
- Patients are discharged from the hospital with a catheter in place to drain urine from the bladder into a bag. Be sure to clean the catheter where it exits your penis twice a day with soap and water.
- Should the catheter fall out or malfunction or is not draining urine, patients should call their urologist. Patients should not allow an emergency department or other non-urology physician to replace or manipulate the catheter.
- Some amount of peri-catheter urine leakage is normal, if it is bothersome an adult diaper can be worn over the catheter.
- Occasionally, minimal bleeding may occur around the catheter or be noticed within the urine. This is common and no treatment is needed. If the is excessive bleeding with large clots or which doesn't look normal, one should contact their urologist.

- Patients are allowed to drive after the urinary catheter removal when they are comfortable doing so.
- Patients can return to work after 2–3 weeks from surgery. If the work entails heavy lifting or exertion one should consult the primary team doctor before resuming work.

19.8 Follow-Up After Surgery [7, 8]

- Follow-up is important to monitor how one has recovered from surgery, to check the general state of health, and to detect possible recurrence of cancer.
- First follow up visit is scheduled at 7–10 days from surgery during which the local surgical site is examined, the urinary catheter is removed, Kegel exercises are taught, and the final histopathology report is discussed.
- The second follow-up visit is at 1 month from surgery with S. PSA (total) levels in the blood. After radical prostatectomy, the S. PSA (total) should fall below 0.2 ng/mL.
- The additional treatment if needed, is decided based on the final histopathology report and S. PSA at 3 month and in some cases PSMA PET CT scan.
- If additional treatment is not needed, then for the first 2 years the following visits are scheduled every 3 months with S. PSA (total) levels. Subsequently at 6 monthly intervals for the next 3 years and then annually thereafter.
- After radical prostatectomy for prostate cancer, routine follow-up lasts around 10 years.

19.9 Managing Incontinence [9]

- After the urinary catheter is removed, most patients will have leakage of urine (incontinence).
- Initially, the leakage may be significant (leakage all the time) and will require a change of diapers multiple times a day.
- Patients are taught exercises that strengthen the pelvic muscles. These are called Kegel exercises and they can be done anytime: when lying down, sitting, standing, or walking.
- These exercises will help in regaining urine control (continence).
- Continence usually returns by 6 months. More than 90% of the patients are diaper free at the end of 1 year.

19.10 Sexual Function

- Men find it difficult to have natural erections after radical prostatectomy. Return of the erectile function depends on the patient's age, pre-operative ability to obtain and maintain an erection, and the extent of the cancer.
- The nerves and blood vessels (neurovascular bundles) that control erection are located on either side of the prostate.
- Sometimes these bundles can be preserved during surgery ("nerve-sparing" radical prostatectomy), thereby maintaining the ability to have an erection.
- In some cases, the neurovascular bundles need to be removed because cancer may extend close to them.
- In patients with impotence post-surgery the management options include oral drugs, penile injections, vacuum pumps, and, rarely, penile implants.
- Since the prostate, seminal vesicles, and vas deferens are removed as part of the procedure, there will be no ejaculate (semen) released.
- Despite the inability to achieve erections, patients are still able to have an orgasm (climax) with stimulation to the penis.

19.11 Conclusion

- Radical prostatectomy is the treatment of choice for patients who have localized intermediate disease T1/2, and favourable high risk disease who have life expectancy of more than 10 years.
- Robotic approach is preferred due to its advantages of being minimal invasive surgery, minimum morbidity, early recovery, early return to work with good functional and oncological outcomes.
- It has psychological advantage that disease is eradicated from the body and patient can live long happily.
- The cost of robot, instruments and disposables are expensive, which has so far restricted its use in Asian countries. However, once cheaper and smaller newer-generation robots become available, their routine use will surely rise

References

1. Huynh LM, Ahlering TE. Robot-assisted radical prostatectomy: a step-by-step guide. J Endourol. 2018;32(S1):S28–32.
2. Hoznek A, Katz R, Gettman M, Salomon L, Antiphon P, de la Taille A, Yiou R, Chopin D, Abbou CC. Laparoscopic and robotic surgical training in urology. Curr Urol Rep. 2003;4(2):130–7.
3. Froehner M, Novotny V, Koch R, Leike S, Twelker L, Wirth MP. Perioperative complications after radical prostatectomy: open versus robot-assisted laparoscopic approach. Urol Int. 2013;90(3):312–5.

4. Weizer AZ, Silverstein AD, Young MD, Vieweg J, Paulson DF, Dahm P. Prospective evaluation of pain medication requirements and recovery after radical perineal prostatectomy. Urology. 2003;62(4):693–7.
5. Levinson AW, Lavery HJ, Ward NT, Su LM, Pavlovich CP. Is a return to baseline sexual function possible? An analysis of sexual function outcomes following laparoscopic radical prostatectomy. World J Urol. 2011;29(1):29–34.
6. Penson DF, Litwin MS. Quality of life after treatment for prostate cancer. Curr Urol Rep. 2003;4(3):185–95.
7. Dogra PN, Javali TD, Singh P, Kumar R, Seth A, Gupta NP. Perioperative outcome of initial 190 cases of robot assisted laparoscopic radical prostatectomy, a single center experience. Indian J Urol. 2012;28:159–63.
8. Gupta NP, Singh P, Nayyar R. Outcomes of robot-assisted radical prostatectomy in men with previous transurethral resection of prostate. BJU Int. 2011;108:1501.
9. Gupta NP, Yadav R, Akpo EE. Continence outcome following Robotic radical prostatectomy: our experience from 150 consecutive patients. Indian J Urol. 2014;30:374–7.

Chapter 20
Focal Therapy for Prostate Cancer

Sudhir Kumar Rawal, Ashish Khanna, and Amitabh Singh

Focal therapy is an alternative between Active Surveillance and Definitive Therapy (NPG)

20.1 Introduction

- The standard treatment options for prostate cancer typically involve either surgical removal of the entire prostate (radical prostatectomy) or administration of radiation therapy to the whole prostate.
- These radical treatments may thus be associated with damage to adjacent critical structures, such as external urinary sphincter (muscle responsible for voluntary urine control) and/or the nerves responsible for rigidity of the penis (erection).
- Consequently, patients might experience problems of urinary control (urinary incontinence) or erection (impotence). Although these issues tend to improve in most men within a year after the procedure, however, 5–15% men may continue to experience long term urinary incontinence and 30–65% previously potent men may complaint of impotence [1–3].

S. K. Rawal (✉)
Genitourinary Oncology, Rajiv Gandhi Cancer Institute and Research Centre, New Delhi, India

A. Khanna · A. Singh
Uro-Oncology and Robotic Surgery, Rajiv Gandhi Cancer Institute and Research Centre, New Delhi, India

N. Gupta (ed.), *A Clinical Guide for Cancer Prostate Patient*,
https://doi.org/10.1007/978-981-97-5357-4_20

20.2 What Is Focal Therapy?

- Focal therapy works by treating/removing only the portion of the organ/gland affected by cancer with a safe margin while sparing the remaining normal healthy tissue.
- This concept is not new and is extensively practiced in treatment of cancers affecting the breast, liver, brain, intestine, kidneys, etc.
- Focal therapy for prostate cancer treats only the area with the most aggressive tumor (referred to as **index lesion**) while leaving the rest of the prostate and its surrounding structures (nerves for erection and sphincter for urinary control) intact (Fig. 20.1).
- Focal therapy has the potential to cure certain men with localized prostate cancer while sparing them of the side effects of whole gland treatment modalities (surgery/radiation).

20.3 How Does Focal Therapy Work?

- Focal therapy relies on the use of energy sources to generate extreme temperatures within the prostate, that lead to tumor destruction.
- The index lesion along with margin of normal prostatic tissue is treated.
- Available options include cryoablation, laser ablation, high-frequency ultrasound (HIFU) ablation and irreversible electroporation (IRE) (Fig. 20.2).
- **Focal cryoablation** uses a needle thin probe to rapidly cool the cancerous tissue to around −40 °C (called as freeze phase). This extreme cooling kills cancer cells. This area is then gradually warmed (called as thaw phase) and then this freeze-thaw cycle is repeated. Two “freeze-thaw” cycles have been shown to be effective.
- **HIFU** converges ultrasound waves at a point in the prostate just like a magnifying glass focuses sunlight. The convergence heats the tumor tissue to around 60 °C and kills cancer cells (Fig. 20.3).

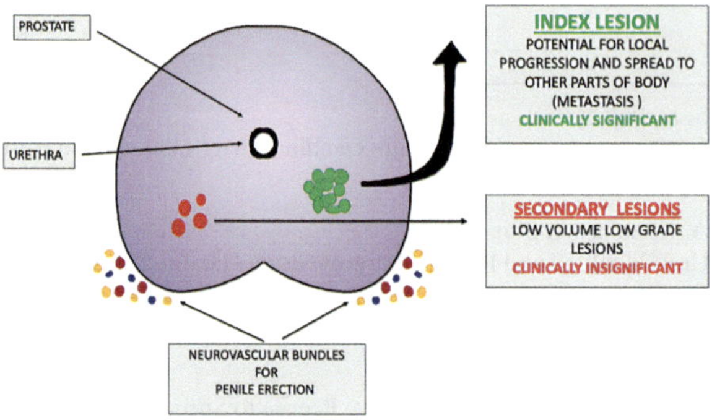

Fig. 20.1 Concept of index lesion as driver of local progression and distant spread (metastasis) of prostate cancer. Secondary lesions with low volume low grade disease seldom progress or spread and therefore may not require active treatment

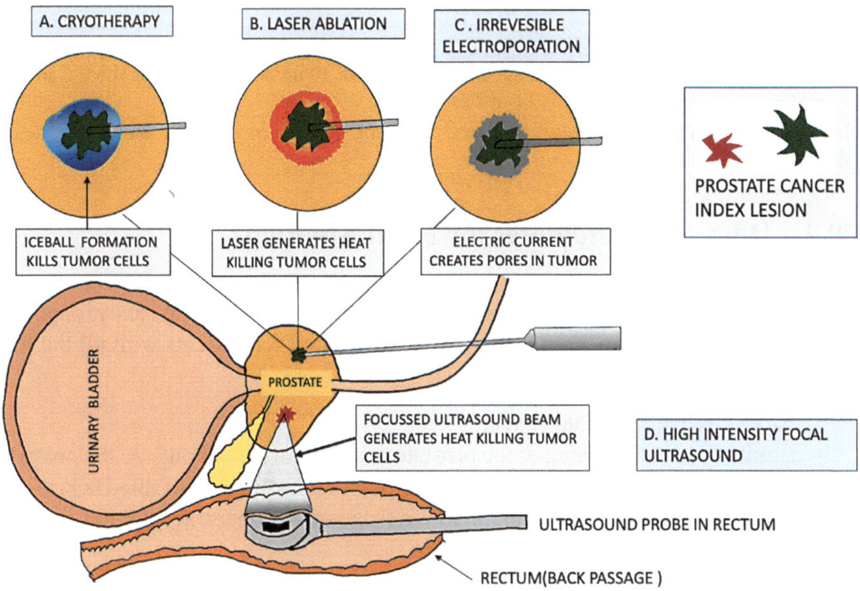

Fig. 20.2 Various options for focal therapy delivery with mechanism of tumor cell destruction

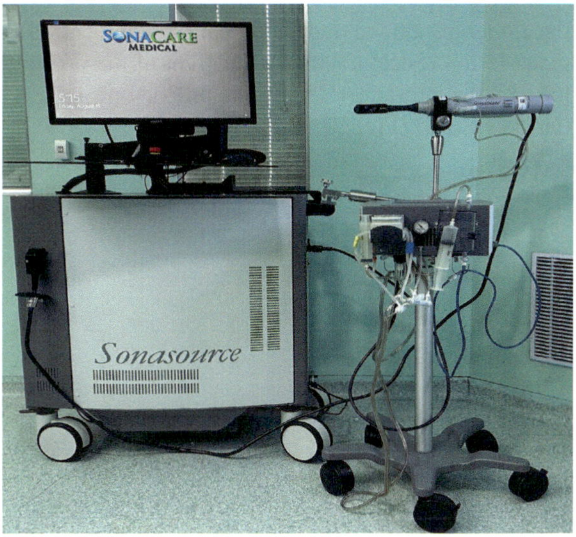

Fig. 20.3 Focal therapy provides minimally invasive, radiation free bloodless ablation of the prostate. The Sonablate system is one such system that delivers high intensity focused ultrasound (HIFU). (Source: Rajiv Gandhi Cancer Institute)

- **IRE** uses one small pulses of electric current to cause holes in cancer cells to kill it and laser ablation uses laser energy to kill cancer cell.

All these modalities rely on real time intraoperative ultrasound to accurately visualise the prostate, localise and ablation specific regions of the prostate.

Table 20.1 Risk stratification according to clinical characteristics

Risk category	Clinical characteristics
Low risk	Prostate specific antigen (PSA) less than 10 ng/mL and Gleason score 6 and stage of cancer is T1 or T2a
Intermediate risk	PSA level is between 10 and 20 ng/mL or Gleason score is 7 and cancer stage is T2b

20.4 Who Are Candidates for Focal Ablation?

- Focal therapy is a treatment option for a select group of prostate cancer patients.
- Although there are no strict criteria, it is best suited for patients with **all** the following **characteristics**:
  - Biopsy proven prostate cancer
  - Limited to only one area of the prostate and has not spread outside the prostate
  - With low to medium risk of spreading outside the prostate (Table 20.1)
  - Some men with small areas of low-risk prostate cancer in addition to their index lesion may also be candidates

20.5 Types of Focal Therapy

Focal therapy can be done according to the location of the index lesion. It can one index lesion (Focal ablation), quadrant of the prostate (Quadrant ablation), or one lobe of the prostate (Hemi ablation) or one lobe + one quadrant of other lobe (Hockey Stick) (Fig. 20.4).

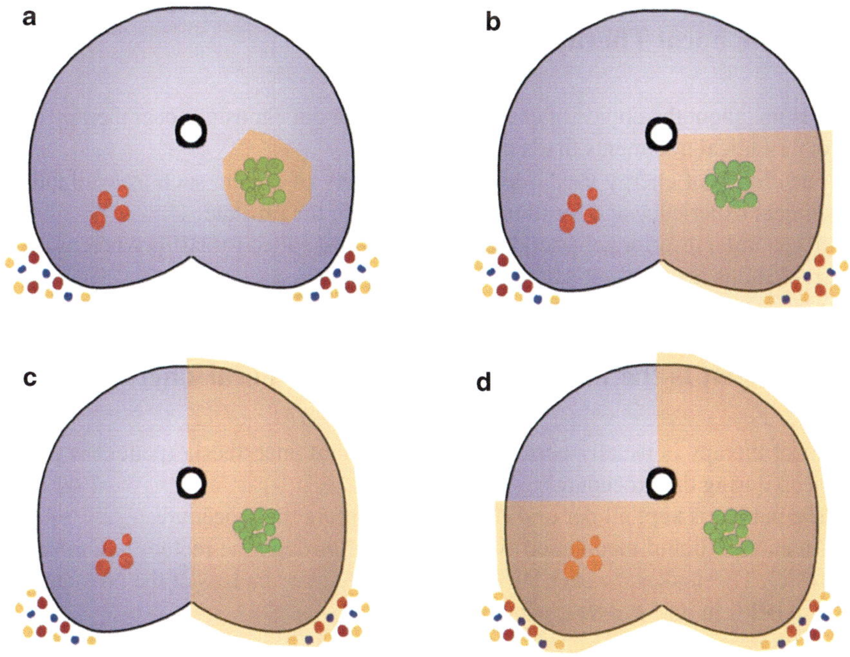

Fig. 20.4 Templates for ablation of prostate cancer. (**a**) Focal lesion ablation, (**b**) quadrant ablation, (**c**) hemi ablation, and (**d**) hockey stick ablation. (Source: van Luijtelaar et al. [4])

20.6 How Does the Clinician Determine Suitability for Focal Therapy?

The urologist would typically determine suitability based on information available from the following tests

- **MRI prostate**: Studies have shown that a special MRI known as multiparametric MRI of the prostate accurately localises areas in the prostate with high probability of harbouring clinically significant cancers. In addition, MRI guides urologists to target the suspicious region during the biopsy.
- **Biopsy**: Prostate biopsy is a must to confirm that cancer is localised to only one part of the prostate with index lesion (Gleason 4).
- **CT chest and abdomen/bone scan/PSMA PET scan**: These tests are done if required to confirm that disease is localised to the prostate and has not spread to other parts of the body. These may not be required in all cases.

20.7 Which Energy Modality Should a Patient Select for Focal Therapy?

- It is not about the choice of energy, rather it is about choosing focal therapy over other radical treatments that's crucial.
- The choice of energy used depends on a variety of factors such as availability, clinical experience and location of lesion within the prostate.
- In general, posterior peripheral lesion may be best suited for HIFU whereas anterior lesions may be best suited for cryotherapy/IRE.

20.8 What Is the Preoperative Process for Focal Therapy?

- Focal therapy is usually performed under general anaesthesia (patient is put to sleep during the procedure)
- The patient is kept nil per oral (NPO) 4–6 h before the procedure.
- An enema (liquid medication) is administered through the rectum to empty the bowel. This makes it easier for the doctor to see clear images of the prostate and also helps in uniform treatment delivery.

20.9 Steps of the Procedure

- A Intravenous Line is placed in one of the veins of the hand through which the anaesthetist shall administer medication to sleep.
- Once under the effect of anaesthesia breathing is maintained by used of highly specialised machines (ventilator)
- A single dose of antibiotic is administered.
- An ultrasound probe is placed from the rectum. This aids in real time visualisation of the prostate gland and tracks progress of procedure.
- Depending on the size of lesion, the procedure can take up to 2.5 h.
- Sometimes, a prostate obstructing the urine flow me need to be resected (procedure called TURP). This is done endoscopically (instrument is passed via urinary passage) without any skin cuts/incisions.
- At the end of the procedure, a soft tube shall be placed via the urinary passage to drain the urine (known as catheter).
- As the effect of anaesthesia weans off, the patient wakes up comfortably and is shifted to the post operative recovery room.

20.10 Steps After the Procedure

- The patient is shifted to room within few hours of the procedure.
- Oral diet is resumed.

- The patient is discharged on the same day or on the next day with the urinary catheter in place.

20.11 What Can One Expect After Discharge?

- Pain and bruising in the region between the testicles and rectum is common after the procedure
- For this, round the clock anti-inflammatory and pain medication are prescribed for a few weeks. These medications are to be taken even in the absence of pain as the help restrict the inflammatory process going on in the treated area of the prostate.
- The Foley catheter is removed in a week's time.
- Burning sensation/Urgency to pass urine can be experienced and resolves over a period of few weeks.

20.12 What Are the Advantages of Focal Therapy?

Focal therapy has the following advantages:

- Usually done as a day care procedure
- Does not require any skin cuts/incision
- Relatively painless procedure
- Recovery is usually quick and most men return to their normal activities within 2–3 weeks
- Treats the small cancer areas with little damage to nearby tissue, nerves and muscles. Consequently, the rates of urinary leakage and erectile dysfunction are significantly lower than radical treatments
- Even after recurrence, the patient remains eligible for repeat session of focal therapy or other options (surgery/radiation)

20.13 What Are the Disadvantages of Focal Therapy?

- Long term cancer survival data (after 10 years) is not yet available for focal therapy.
- About 20–30% patients may need a second session of focal therapy for recurrent cancer.
- Less than 1% patients may develop a hole between the rectum and the urinary passage (rectourethral fistula). This may need urethral catheter placement for few months or surgical repair in case it does not heal on its own [5].

20.14 How Is the Follow Up After Focal Therapy?

- After focal therapy, the patients PSA level is expected to fall and then remain low.
- PSA testing is performed at 3–6 monthly interval for the first few years after the procedure.
- The urologist may perform imaging in the form of MP MRI at frequent intervals to track response of the lesion to treatment, development of new lesions in the treated area or elsewhere in the prostate.
- A rising PSA may indicate that the prostate cancer has returned. This would require repeat MRI of the prostate and if required a repeat biopsy.

20.15 What to Do If Cancer Recurs After Focal Therapy?

- Multiple options exist for recurrence after
  - Second session of focal therapy
  - Radiation therapy
  - Surgery: Radical prostatectomy
  - Hormonal therapy
- The best option needs to individualised and is taken after thorough discussion between the patient and clinician.

References

1. Ficarra V, Novara G, Ahlering TE, Costello A, Eastham JA, Graefen M, Guazzoni G, Menon M, Mottrie A, Patel VR, Van der Poel H. Systematic review and meta-analysis of studies reporting potency rates after robot-assisted radical prostatectomy. Eur Urol. 2012;62(3):418–30.
2. Ficarra V, Novara G, Rosen RC, Artibani W, Carroll PR, Costello A, Menon M, Montorsi F, Patel VR, Stolzenburg JU, Van der Poel H. Systematic review and meta-analysis of studies reporting urinary continence recovery after robot-assisted radical prostatectomy. Eur Urol. 2012;62(3):405–17.
3. Stanford JL, Feng Z, Hamilton AS, Gilliland FD, Stephenson RA, Eley JW, Albertsen PC, Harlan LC, Potosky AL. Urinary and sexual function after radical prostatectomy for clinically localized prostate cancer: the Prostate Cancer Outcomes Study. JAMA. 2000;283(3):354–60.
4. van Luijtelaar A, Greenwood BM, Ahmed HU, Barqawi AB, Barret E, Bomers JGR, Brausi MA, Choyke PL, Cooperberg MR, Eggener S, Feller JF, Frauscher F, George AK, Hindley RG, Jenniskens SFM, Klotz L, Kovacs G, Lindner U, Loeb S, Margolis DJ, Marks LS, May S, Mcclure TD, Montironi R, Nour SG, Oto A, Polascik TJ, Rastinehad AR, De Reyke TM, Reijnen JS, de la Rosette JJMCH, Sedelaar JPM, Sperling DS, Walser EM, Ward JF, Villers A, Ghai S, Fütterer JJ. Focal laser ablation as clinical treatment of prostate cancer: report from a Delphi consensus project. World J Urol. 2019;37(10):2147–53. https://doi.org/10.1007/s00345-019-02636-7. Epub 2019 Jan 22. PMID: 30671638; PMCID: PMC6763411
5. Netsch C, Bach T, Gross E, Gross AJ. Rectourethral fistula after high-intensity focused ultrasound therapy for prostate cancer and its surgical management. Urology. 2011;77(4):999–1004.

Chapter 21
Radiation Therapy for Prostate Cancer

Tejinder Kataria, Susovan Banerjee, and K. P. Venkatesan

21.1 Introduction

- Prostate cancer incidence, as per Globocan, 2020 is 14.1%. Prostate cancer has become a major health problem in industrialized world during the last decades of the twentieth century contributing to three fourth of the registered cases across the globe.
- In India with increased migration of rural population to the urban areas, changing life styles, increased awareness, and easy access to medical facility, more cases of prostate cancer are being diagnosed. Current incidence in India is 10.66 per 100,000/year [1, 2].
- Treatment option for carcinoma prostate depend upon the stage and risk group category (details in other chapter).
- For very low and low risk category active surveillance policy is recommended, however the patient must be kept on close monitoring under a trained urologist [3].
- Patients with organ confined prostate cancer viz. T1, T2N0M0 and intermediate risk disease group category can be treated with either only surgery or radiation therapy supplemented with a short course androgen (hormone) ablation of

T. Kataria (✉)
Division of Radiation Oncology, Medanta Cancer Institute, Medanta, Gurugram, India
e-mail: tejinder.kataria@medanta.org

S. Banerjee
Division of Radiation Oncology, Medanta, Gurugram, India
e-mail: susovan.banerjee@medanta.org

K. P. Venkatesan
Division of Radiation Oncology, Medanta Cancer Institute, Medanta, Gurugram, India

Radiation Oncology, Apollo Speciality Hospital, Vanagaram, Chennai, India
http://www.apollohospitals.com

N. Gupta (ed.), *A Clinical Guide for Cancer Prostate Patient*,
https://doi.org/10.1007/978-981-97-5357-4_21

6 months. The results are equivalent at 10 years, post diagnosis provided there are no other life shortening co-morbidities like end stage heart or renal diseases [4].
- Patients with stage T3N0M0, T3N1M0, T4N0M0, T4N1M0 are high risk and very high-risk category patients. They are candidates for neo-adjuvant or upfront total androgen ablation therapy often with two drug combination or bilateral orchiectomy if more than 70 years of age followed by external beam radiation and/or brachytherapy/stereotactic body radiosurgery (SBRT) boost [5, 6].
- Other high risk and very high-risk categories include T2N1M0, T3/T4N0/N1M1 patients with a Gleason's score of >8–10 and serum PSA >20 ng/mL and need to be stratified for upfront total androgen ablation (often supplemented with newer hormonal agents like Abiratarone and Enzulatamide) followed by radiation to the primary prostate, lymph nodes and/or metastatic disease depending upon oligo or poly metastatic disease [7].

21.2 History of Radiation for Prostate Cancer

- The first reports of prostate cancer treatment with radiation were published at the beginning of twentieth century with application of Radium sources within or close to prostate in palliative setting as an alternative to surgery [8]. The techniques placed Radium sources through urethra or rectum and were uncomfortable for the patients, hence they were not widely adopted.
- In 1970s Whitmore et al. standardized the open surgical approach for application of radioactive Iodine seeds into the prostate, thus reviving an interest in treatment of prostate under direct visualization, that resulted better disease staging and dose distribution [9]. However, Holm et al. reported the placement of radioactive seeds through trans perineal route under transrectal ultrasound guidance and since then Iodine seeds have become a monotherapy for early prostate cancer [10].
- External beam radiation was explored in treatment of prostate cancer in 1940s by Del Regato's group and in 1950s by Macolm Bagshaw. These studies established the radio curability of prostate cancer [11, 12] using a rotational platform for the patient with megavoltage radiation unit. Since then, long strides have been taken in technological improvement in imaging, staging and role of radiation in management of various stages of cancer prostate.

Indications of radiation for prostate cancer are well defined for radical, adjuvant and palliative settings.

- Intent with Indication of Radiotherapy for carcinoma Prostate
- Radical Radiation with or without hormone ablation
  - T1N0M0, T2N0M0, T2N1M0, T3N0M0, T3N1M0, T4aN0M0, T4N1M0

- Radiation is the treatment of choice for elderly frail persons with comorbid conditions and patient's refusal to surgery.
- Post op Radiation therapy:
  - pT3N1M0; T4N0M0; T4N1M0 pathological stage
  - Any patient with close or positive margin
  - Presence of extracapsular spread in pathology specimen
  - Multiple positive nodes in final specimen
  - R1 or R2 resection or preop PSA >10 ng/mL and PSA velocity >2 ng/mL/year
  - As early salvage radiotherapy with detection of early PSA relapse (<0.2 ng/mL)
  - Post Radical Prostatectomy recurrent disease or biochemical relapse
  - Palliative Radiation for painful bony or brain metastases
  - Salvage RT/Re RT

21.2.1 Radical RT Techniques

- External Beam Radio Therapy
  - Conventional radiatiation 180–200 cGy/day for 5 days/week over 7–8 weeks
  - Hypofractionation with Robotic Radiosurgery platform Cyberknife or Linear accelerator/Stereotactic Body Radiotherapy

    Moderate hypofractionation (220–240 cGy/fraction) delivered over 20–28 fractions, 5 days/week. Extreme hypofractionation (600–700 cGy/fraction) delivered over 5–7 fractions, 3 fractions/week on alternate day.
- EBRT + Brachytherapy Boost (interstitial) or External radiotherapy with conventional fractionation and stereotactic body radiotherapy boost with extreme hypofractionation using Robotic Radiosurgery platform Cyberknife.
  - Temporary implant—HDR (Iridium)

    Permanent implant—I-125
- Protons (charged particles) and Photons (high energy X-Rays) are the modalities of radiation energy used for curing prostate and other cancers and are delivered through specific machines with remote control and extensive planning per patient.

21.2.2 Simulation for Radiotherapy

- Mould room-patient is positioned in supine condition and an immobilization device like pelvic cast or Vacuum cushion from abdomen to mid thigh is made with tegaderm marking. Planning CT scan is taken from umbilicus (D12) to mid thigh with bladder protocol.

21.2.3 Radiotherapy Doses

- For EBRT (Minimum standard is Intensity Modulated Radiotherapy preferably with daily imaging):
  - Radical doses—70–78 Gy with 2.5 or 2 Gy per fractions
  - Post op doses—60–66 Gy with 2 Gy/1.8 Gy per fraction
  - CK/SBRT—5–12 Gy in 3–5 fractions
  - Simultaneous integrated nodal boost (for gross nodes) up to 54–64 Gy when possible
  - Palliative RT—30 Gy/10 fractions or 20 Gy in 4–5 fractions
- For Brachytherapy most common schedules are:
  - External beam radiotherapy (EBRT) is delivered to a dose of 46–50 Gy @ 2.0 Gy/fraction to the prostate and pelvic lymph nodes in T3 and beyond or high risk prostate cancer. This is followed by a boost to the prostate alone with a dose of 26–30 Gy @ 2.0 Gy/fraction
  - Brachytherapy—LDR 20–30 Gy in 2–3 fractions. HDR = 9.5 Gy × 2 fraction/12–15 Gy in 1 fraction

21.3 Techniques of Radiation Therapy

21.3.1 Brachytherapy

In prostate cancer treatment by radiotherapy, brachytherapy can be integral part of the treatment. Clinical results of prostate brachytherapy are comparable with other modalities of radiotherapy with comparable toxicities [13]. For low and intermediate risk cancer patients it may be the sole modality of treatment [14]. For high risk and locally advanced prostate cancer patients, using brachytherapy as boost before or after external beam radiotherapy gives superior clinical results in terms of biochemical progression-free survival (BFS) [15].

- Interstitial brachytherapy to prostate can be low dose rate (LDR) or high dose rate (HDR). In LDR permanent seed implantation technique I125 or Palladium-103 (Pd-103) are commonly used [16]. Seeds with very low dose radioactivity activity are left inside the prostate permanently, the procedure is trans rectal ultrasound (TRUS) guided and is generally proceeded by a preplanning and followed with post procedure verification imaging. HDR prostate brachytherapy involves placing temporary catheters inside the prostate in the operating theater under TRUS guidance, then dose optimization is done with help of dedicated software on computer tomography (CT) or ultrasonography (USG) imaging. Modern USG based planning systems allow treatment planning on a real time basis within the operating theatre and treatment thereafter without moving or changing position of the patient, so that there is no chance of catheter displacement and planning alteration [17]. In HDR brachytherapy after treatment delivery the catheters are removed from the body.

- Historically LDR seed implant has been most popular and was considered gold standard previously, but both HDR and LDR treatment shows similar results [13].
- The benefits of HDR-BT over LDR-BT is using the same source for other cancers, and has a lower operator dependence, almost no radiation exposure to health care professionals and better options of plan optimization. In recent times popularity of HDR prostate brachytherapy has increased worldwide due to such advantages and it is also practiced in some centers of India, including Medanta Gurgaon.

Figure 21.1 shows an Integrated Brachytherapy Unit at Medanta along with the High Dose Rate (HDR) MicroSelectron system.

Figure 21.2 shows the sagittal, transverse, and coronal sections with a three-dimensional cartoon of the catheters for HDR brachytherapy and the anatomical separation of bladder and rectum from high dose spill.

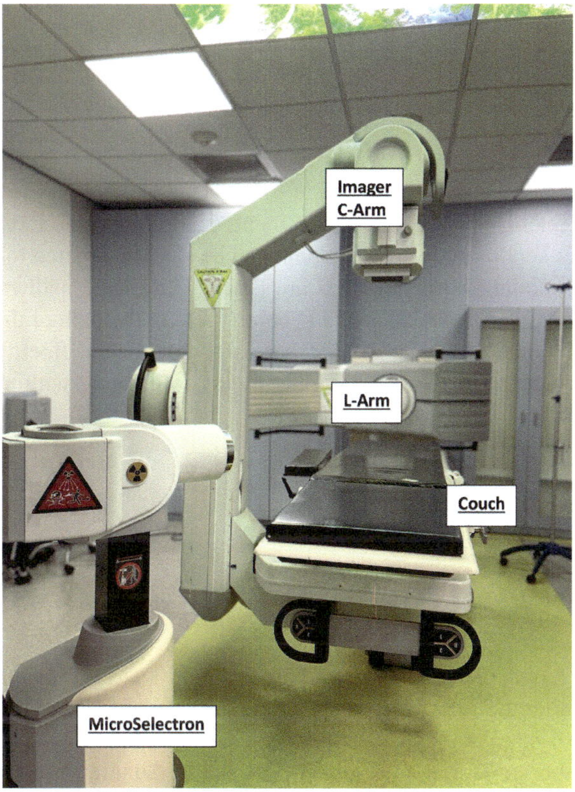

Fig. 21.1 High Dose Rate MicroSelectron (Elekta registered)-Remote controlled, Integrated Brachytherapy unit for treatment verification. (Courtesy: Division of Radiation Oncology, Medanta)

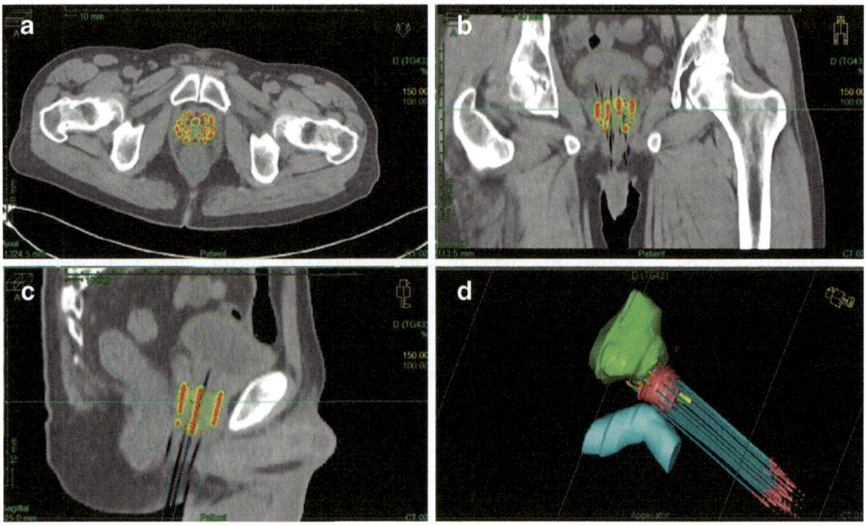

Fig. 21.2 High dose rate brachytherapy: (**a**) Axial section Ct scan with needle reconstruction and sparing of urethra (yellow), rectum (blue). (**b**) Sagittal CT section with longitudinal display of catheters. (**c**) Coronal Ct section with catheters in the prostate. (**d**) 3-D reconstruction of catheters in prostate, sparing of bladder (green) and rectum (blue). (Courtesy: Division of Radiation Oncology, Medanta)

21.3.2 Teletherapy

Tele(distant); therapy-(treatment)-is the technique in which the source of radiation is kept away from the patient and beams are directed to the area of interest. The surrounding normal tissues are blocked from the path of the radiation by either (a) creating lead blocks and keeping them in the path of the radiation beam, (b) having automated blocks of the beam by half beam block attachment to the radiation machine, (c) introducing multi leaf collimators within the beam path in the linear accelerator.

21.4 Radiation Therapy

21.4.1 2-D or Two-Dimensional Radiation Therapy

2-D or 2-dimensional radiation therapy was the technique deployed in the 1950s after the megavoltage systems were available for clinical use. Here a beam energy of more than 1 MeV (million electron volts) was supplied by the machine. Before the cobalt-60 machines megavoltage radiation was produced by Van de Graaff

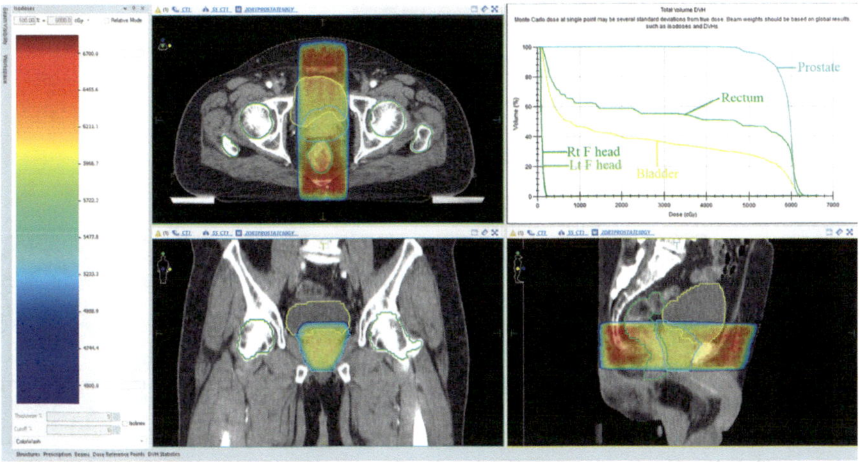

Fig. 21.3 Antero-posterior deployment of two radiation beams in 2-D technique delivered radiation dose to rectum and bladder leading to radiation proctitis and cystitis in 20–25% of cases. The 2-D treatments were given in 1960s–70s (Virtual dose profile created on a planning system and not delivered to any patient). (Courtesy: Division of Radiation Oncology, Medanta)

generator and bevatrons that were built in 1930s [18]. The megavoltage beams could be used to treat deeper tumours like prostate cancer, however, the beams could be directed through limited directions only and were generally open beams. Dose was prescribed to some points and fluoroscopic images were used for planning and treatment verification. Such treatments resulted in delivery of wider exposure of normal organs and delivery of higher doses to surrounding structures like bladder, rectum and pelvic bones while treating prostate cancer (Fig. 21.3).

21.4.2 3-D Conformal Radiotherapy (3-D CRT)

This is a technique of radiotherapy in which volumetric images were used for target/organ delineation and radiotherapy planning. The radiation beams are delivered either from one anterior or two posterior oblique portals or one beam each from antero-posterior and one beam each from lateral portals are deployed. The dose is divided into 3 or 4 beams to reduce the high dose regions from rectum, bladder and other organs at risk of receiving radiation. The 3D CRT configuration was used in the 1980s with creation of lead blocks kept on a tray within the radiation beam path on Cobalt-60 and early Linear accelerator machines. With the advent of multi-leaf collimators external lead blocks were removed and addition of CT-Planning images made it possible for radiation therapy to move into an era of intensity modulated radiotherapy in 1990s [19, 20] (Fig. 21.4).

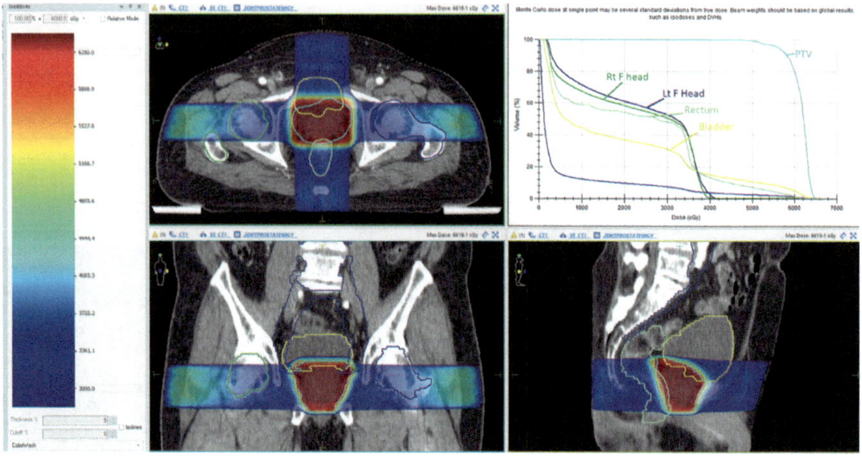

Fig. 21.4 3D conformal radiotherapy plan and beam arrangement for prostate cancer with reduction in bladder and rectal dose due to the spatial distribution of dose. (Courtesy: Division of Radiation Oncology, Medanta)

21.4.3 Intensity Modulated Radiotherapy (IMRT)

Intensity modulated radiotherapy is an advanced technique over volumetric radiotherapy planning by 3D CRT, where the treatment beams are inversely planned to be delivered to the prostate and pelvic lymph nodes (if required) in the patient respecting the tolerance of rectum, bladder, pelvic bones and small intestines. The patient is immobilized in an orfit cast (a plastic mould) or in a vacuum cushion to minimize external body motion and daily treatment reproducibility. With approximately 350–500 mL of urine holding in the bladder, the patient undergoes a pelvic, CT-Scan preferably with IV contrast. The patients orfit cast and board or the vacuum cushion are placed on a flat tabletop mounted on the CT table with the patient lying over that. The planning CT scan is done with thin sections to permit anatomical delineation of prostate, pelvic nodes, pelvic blood vessels, bladder, rectum, bones, and intestines. After the Radiation Oncologist delineates all the anatomical structures she/he will prescribe the tumoricidal dose to the targets (regions of interest) as well as advice limit of exposure for the organs at risk (normal tissues to be spared). Then a dedicated high capacity computer planning system and software programme is used to prepare a dose map respecting or avoiding the normal tissues and delivering the curative/sterilizing dose to the prostate/pelvic nodes. The final dose map is converted to a deliverable beam configuration, which is shaped by the multi-leaf collimators of the Linear Accelerator or Tomotherapy machine with segmentation and transferred to delivery console of the treatment machine. The total dose is divided into number of fractions ranging from 20 to 38, 1 fraction per day and 5 fractions are delivered per week (Fig. 21.7).

21.4.4 Volumetric Modulated Arc Therapy (VMAT)

VMAT is an advanced radiation therapy planning and delivery technique using 360 beam configurations deployed either in single plane or non-co-planar method to conform the dose envelop to the cancer and protect the normal tissue complication [21]. VMAT also allows for simultaneous integrated boost (SB) to be planned for the dominant nodule within the prostate or an enlarged lymph node that needs extra dose. In this technique not only do the gantry and the multi-leaf collimators configure the shape, but the beam current is also altered to reduce the dose deposition in normal organs as determined by biological algorithms to minimize any side effects. Figure 21.8 shows the dose plan for a 360 VMAT plan for prostate cancer.

Figs. 21.5 (manufacturer Elekta BV) and 21.6 (Tomotherapy HD (manufacturer Accuray) show the two types of radiotherapy machines deployed for prostate teletherapy to deliver IMRT with a 7 beam arrangement on Linear accelerator with organ sparing; 360° VMAT (Volumetric Modulated Radio Therapy) on a Linear accelerator and a planand 64 field IMRT with Tomotherapy respectively in Figs. 21.7, 21.8 and 21.9.

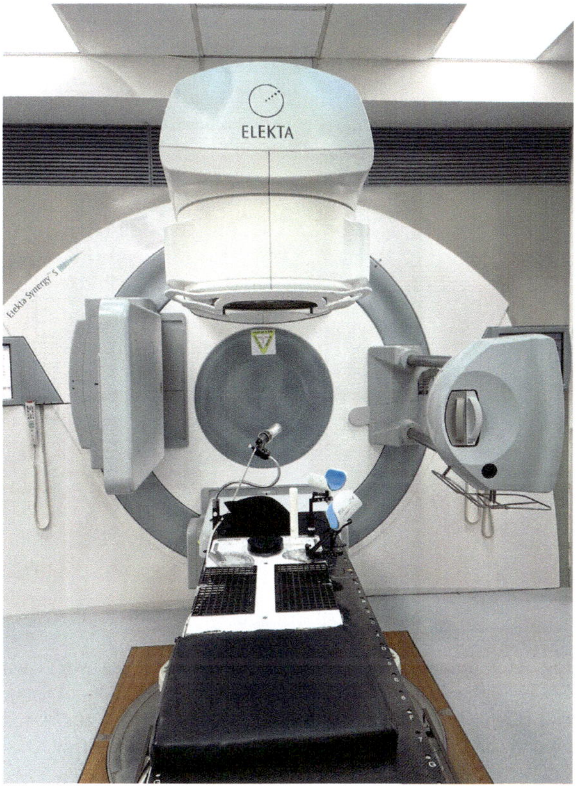

Fig. 21.5 Linear accelerator enabled to deliver IMRT, VMAT and IGRT. (Courtesy: Division of Radiation Oncology, Medanta, Gurgaon)

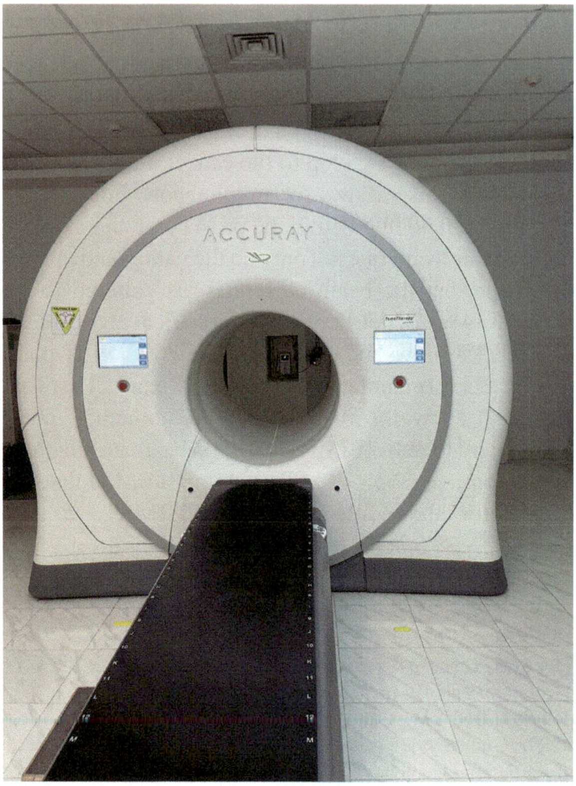

Fig. 21.6 Tomotherapy HD-Ring gantry with CT and translational couch enabled for IMRT and IGRT integrated delivery system. Should will remove the Breast board ABC set up? (Courtesy: Division of Radiation Oncology, Medanta, Gurgaon)

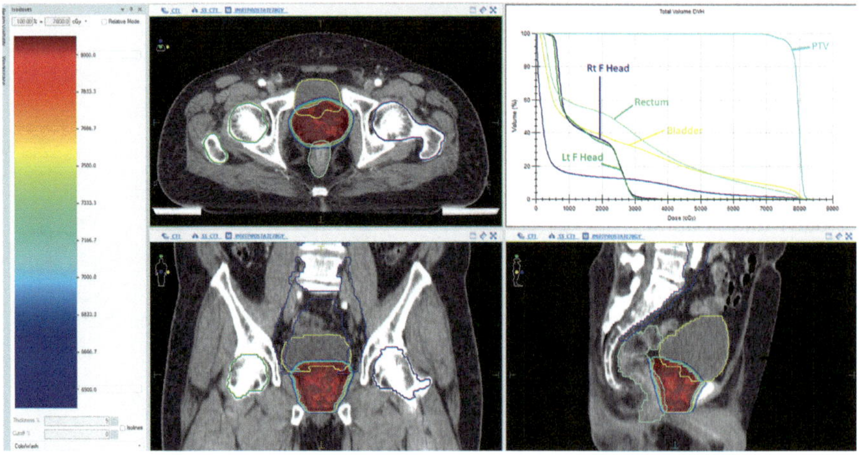

Fig. 21.7 Intensity Modulated Radiation Therapy (IMRT) with 7-beams to deliver a dose of 78 Gy/39 fractions to prostate, reducing the dose to 50% of rectum to less than 25 Gy and less than 800 cGy to bladder. The reduction in dose to organs at risk reduces the probability of radiation proctitis and radiation cystitis to less than 2–5%. (Courtesy: Dr. Tejinder Kataria)

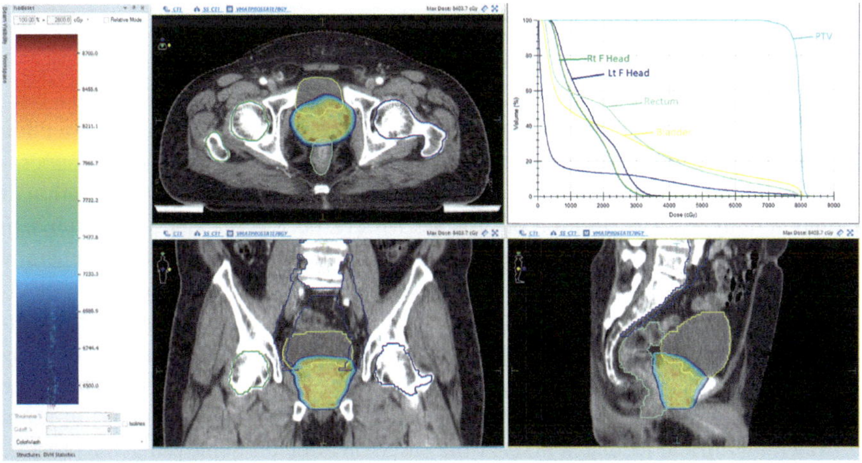

Fig. 21.8 Volumetric Modulated Arc Therapy using 360 beam configurations with sparing of rectum, bladder, pelvic bones, femoral heads, and intestines delivered on a Linear Accelerator. (Courtesy: Division of Radiation Oncology, Medanta)

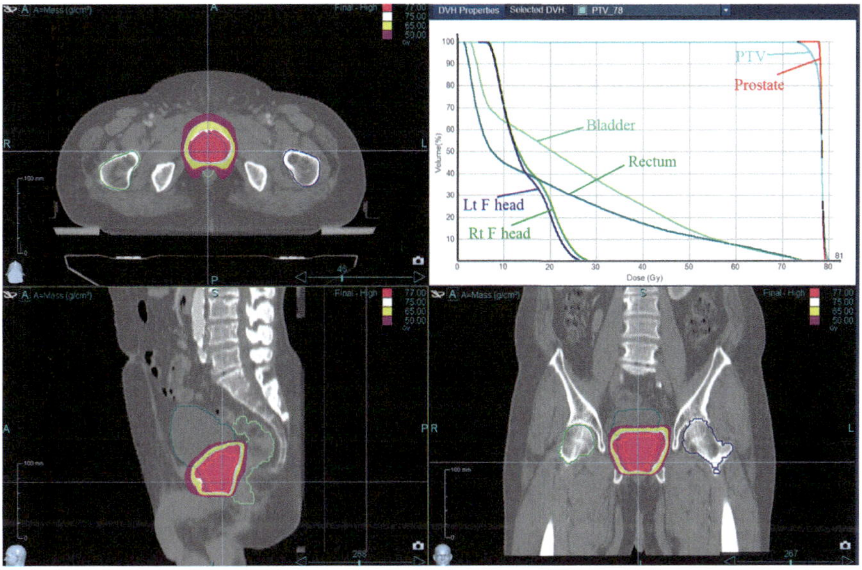

Fig. 21.9 Tomotherapy plan with 64 beams using IMRT to deliver 78 Gy/39 fractions to prostate. (Courtesy: Division of Radiation Oncology, Medanta)

21.4.5 Image Guided Radiotherapy

The prostate gland has its own deformation that occurs during the delivery of a fraction of radiation as well as it is subjected to movement in antero-posterior direction by the rectum filling with either fecal matter or gas. It is also subject to supero-inferior motion due to bladder filling, that occurs constantly through the duration of radiation delivery. The position of prostate with respect to bladder and rectum can be determined by in-treatment room imaging known as Image Guided Radiotherapy devices. These include (a) On-board imaging to create a virtual CT-Scan as is available in present day Linear accelerators, (b) Megavoltage-imaging using the treatment beam to create an image of treatment area, (c) In-built KV CT scan-Tomotherapy HD equipment has it as well as the couch translates as a CT couch to take real time images of target and OARs, (d) CT-on rails, (e) MR-Linear accelerator, (f) In-room X-Ray imaging and feedback in real time based on fiducials deployed in the prostate gland for Robotic Radiotherapy-Cyberknife Radiosurgery system. Image Guided Radio Therapy (IGRT) allows a reduction in margins to the prostate and save the undue dose deposition within the rectum and bladder thereby eliminating the acute and long-term side effects of treatment [22].

21.4.6 Cyberknife Robotic Radio Surgery

The conventional Linear Accelerator weighs around 1770 kg; the Cyberknife linear accelerator is miniaturized to 24 inches. and weighs about 250 kg. It is mounted on the Kuka robot to enable 6 degrees of freedom of the robot, coupled to the 270 degrees of rotation of the couch, >1200 beam configurations available with the multi-leaf collimator within the head of the Linear accelerator empowering the system to create a sculpted dose to the target and region of interest. The margin given to prostate on CT imaging done at 1 mm sections, is 2 mm and gold seeds are placed within the prostate to track its rotations along 3 axis and also, antero-posterior, medio-lateral and supero-inferior transitional movement in real time. In case of prostate moving out of 2 mm margin the Robot will not deliver radiation. The Robot is guided by an In-room imaging device that tracks the movement of prostate varying from 3 to 30 s creating a Real-time motion feedback loop to the robotic delivery system [23].

Figure 21.10 shows the Upgraded Cyberknife-VSI system available at Medanta, Gurgaon.

Figure 21.11 shows the dose distribution for a 7-fraction radical dose delivery with Cyberknife for prostate cancer.

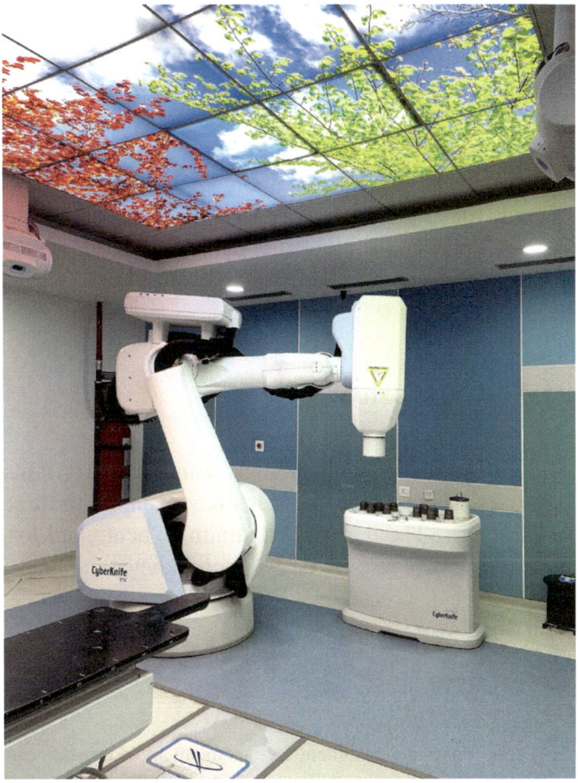

Fig. 21.10 Cyberknife-VSI (Accuray Registered) Robotic Radio Surgery system. (Courtesy: Division of Radiation Oncology, Medanta)

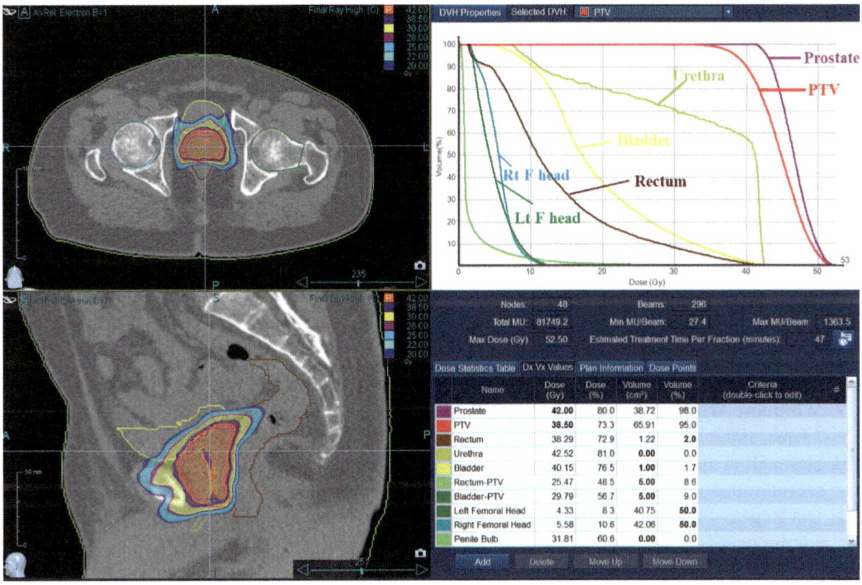

Fig. 21.11 Spatial dose distribution in a prostate cancer with Cyberknife robotic System creating sharp fall of dose in OARs. (Courtesy: Division of Radiation Oncology, Medanta)

21.4.7 *Protons*

Protons are a form of teletherapy using high energy charged particles like hydrogen atoms (protons) or carbon ions to deposit radiation within the target. The heavy charged ions can be programmed to deposit almost all their energy in the target as they follow the principle of Bragg Peak Effect and can deliver all the dose at the end of their track. Last 10 years have seen an increasing use of protons for prostate cancer, however, the clinical results are equivalent with photon teletherapy albeit at a very high financial cost to the patient [24].

21.5 Conclusion

The advancements in technology, computer algorithms and experimental radiobiology over the last three decades have shaped the treatment of prostate cancer in radiotherapy. The intent is to minimize acute and long-term morbidity and maximize the control rates in early, intermediate, and high-risk prostate cancers.

References

1. Draft consensus document for management of prostate cancer. https://main.icmr.nic.in/sites/default/files/whats_new/Prostate_cancer_draft_document01092022.pdf.
2. https://ncdirindia.org/All_Reports/Report_2020/Factsheet/Fact_Sheet_2020.pdf.
3. Bill-Axelson A, et al. Radical prostatectomy or watchful waiting in prostate cancer—29-year follow-up. N Engl J Med. 2018;379:2319–29.
4. Greene DE, Mayadev JS, Valicenti RK. Radiation treatment for patients with intermediate-risk prostate cancer. Ther Adv Urol. 2012;4(3):113–24. https://doi.org/10.1177/1756287212442977.
5. Morton GC. High-dose-rate brachytherapy boost for prostate cancer: rationale and technique. J Contemp Brachytherapy. 2014;6(3):323–30. https://doi.org/10.5114/jcb.2014.45759. Epub 2014 Oct 6.
6. Wang S-C. Whole pelvic radiotherapy with stereotactic body radiotherapy boost vs. conventionally fractionated radiotherapy for patients with high or very high-risk prostate cancer. Front Oncol. 2020;10:814. https://doi.org/10.3389/fonc.2020.00814.
7. Burgess L. A review on the current treatment paradigm in high-risk prostate cancer. Cancers (Basel). 2021;13(17):4257. Published online 2021 Aug 24. https://doi.org/10.3390/cancers13174257.
8. Young HH. Use of radium in cancer of the prostate and bladder. JAMA. 1917;68:1174–7.
9. Whitmore W, et al. Retropubic implantation of iodine-125 in the treatment of prostate cancer. J Urol. 1972;108:918–20.
10. Holm H, et al. Trans perineal iodine-125 seed implantation in prostate cancer guided by transrectal ultrasonography. J Urol. 1983;130:283–6.
11. Del Regato JA. Radiotherapy in the conservative treatment of operable and locally inoperable carcinoma of the prostate. Radiology. 1967;88:761–6.

12. Bagshaw MA, Kaplan HS, Sagerman RH. Linear accelerator super voltage radiotherapy. VII. Carcinoma of the prostate. Radiology. 1965;85:121–9.
13. Zaorsky NG, Davis BJ, Nguyen PL, Showalter TN, Hoskin PJ, Yoshioka Y, Morton GC, Horwitz EM. The evolution of brachytherapy for prostate cancer. Nat Rev Urol. 2017;14(7):415–39. https://doi.org/10.1038/nrurol.2017.76.
14. Schaeffer EM, Srinivas S, Adra N, An Y, Barocas D, Bitting R, et al. NCCN Guidelines® Insights: prostate cancer, version 1.2023. J Natl Compr Cancer Netw. 2022;20(12):1288–98. https://doi.org/10.6004/jnccn.2022.0063.
15. Kee DLC, Gal J, Falk AT, Schiappa R, Chand ME, Gautier M, Doyen J, Hannoun-Levi JM. Brachytherapy versus external beam radiotherapy boost for prostate cancer: systematic review with meta-analysis of randomized trials. Cancer Treat Rev. 2018;70:265–71. https://doi.org/10.1016/j.ctrv.2018.10.004. Epub 2018 Oct 11.
16. Banerjee S, Kataria T, Goyal S, et al. Low dose rate permanent seed brachytherapy: tracing its evolution and current status. Precision Radiat Oncol. 2020;4:89–98. https://doi.org/10.1002/pro6.1096.
17. Morton GC. High-dose-rate brachytherapy boost for prostate cancer: rationale and technique. J Contemp Brachytherapy. 2014;6(3):323–30. https://doi.org/10.5114/jcb.2014.45759. Epub 2014.
18. Halperin EC, Perez CA, Brady LW. Perez and Brady's principles and practice of radiation oncology. 5th ed. Philadelphia: Wolters Kluwer Health/Lippincott Williams & Wilkins; 2008. p. 150. ISBN: 9780781763691
19. Zelefsky MJ, Leibel SA, Gaudin PB, et al. Dose escalation study with three-dimensional conformal radiation therapy affects the outcome in prostate cancer. Int J Radiat Oncol Biol Phys. 1998;41:491–500.
20. Zelefsky MJ, Fuks Z, Harpersett L, et al. Clinical experience with intensity modulated radiation therapy (IMRT) in prostate cancer. Radiother Oncol. 2000;55:241–9.
21. Kuang Y, Wu L, Hirata E, Miyazaki K, Sato M, Kwee SA. Volumetric modulated arc therapy planning for primary prostate cancer with selective intraprostatic boost determined by 18F-choline PET/CT. Int J Radiat Oncol Biol Phys. 2015;91(5):1017–25. https://doi.org/10.1016/j.ijrobp.2014.12.052.
22. Goyal S, Kataria T. Image guidance in radiation therapy: techniques and applications. Radiol Res Pract. 2014;2014:705604. https://doi.org/10.1155/2014/705604. Epub 2014 Dec 17.
23. Bryant CM, Henderson RH, Nichols RC, Mendenhall WM, Hoppe BS, Vargas CE, Daniels TB, Choo CR, Parikh RR, Giap H, Slater JD, Vapiwala N, Barrett W, Nanda A, Mishra MV, Choi S, Liao JJ, Mendenhall NP, Genitourinary Subcommittee of the Particle Therapy Co-Operative Group. Consensus statement on proton therapy for prostate cancer. Int J Part Ther. 2021;8(2):1–16. https://doi.org/10.14338/IJPT-20-00031.1.
24. Miszczyk L, Namysł Kaletka A, Napieralska A, Woźniak G, Stąpór Fudzińska MA, Głowacki G, Tukiendorf A. Cyberknife radioablation of prostate cancer—preliminary results for 400 patients. Asian Pac J Cancer Prev. 2017;18(4):1007–13. https://doi.org/10.22034/APJCP.2017.18.4.1007.

Chapter 22
Hormonal Therapy in Prostate Cancer

Gopal Sharma

22.1 Introduction

Androgen deprivation therapy commonly known as hormonal therapy prevents testosterone (male hormone) from imparting its effect on prostate cells either by reducing its production or preventing it from reaching prostate cells [1]. It is a commonly used therapy for prostate cancer as these cancer cells rely on testosterone for growth. Therefore, in the absence of testosterone cancer cells inside the prostate either die or grow slowly [1, 2]. A state of testosterone deprivation can be created by hormonal therapy in the form of oral or injectable medicines or by surgical removal of the testis through a procedure called bilateral orchidectomy [1]. Often, hormonal therapy is clubbed with other treatment modalities such as radiation therapy to make it more effective.

22.2 What Are Male Sex Hormones?

Testosterone is the chief sex hormone in men produced primarily in the testes. Some small amount of testosterone is also made in a gland above the kidneys called adrenal gland. Hormones are messengers that carry signals from one part of the body to other. Testosterone (male sex hormone) is responsible for sex drive, making sperms, muscle mass and bone density. Prostate cancer cells also thrive on testosterone, therefore blocking this hormone has been used as a treatment modality for prostate cancer.

G. Sharma (✉)
Urologic Oncology and Robotic Surgery, Medanta Hospital, Gurugram, India

N. Gupta (ed.), *A Clinical Guide for Cancer Prostate Patient*,
https://doi.org/10.1007/978-981-97-5357-4_22

22.3 Source of Male Sex Hormones

In men, testosterone is primarily produced by the Leydig cells in the testes. However, small quantities of this hormone is also produced by the adrenal glands.

22.3.1 Mechanism of Action of Hormonal Therapy

Stimulation for testosterone hormone production begins in hypothalamus and pituitary gland (in brain). Lieutenising hormone-releasing hormone (LHRH) and Lieutenising hormone (LH) are produced by hypothalamus and pituitary gland respectively. LH acts on Leydig cells in the testis and cause increased production of testosterone. Testosterone hormone binds to its receptors on tissues such as prostate cells. This receptor-hormone complex is then translocated to nucleus where it causes its effects on genes ultimately leading to increased prostatic tissue. Figure 22.1 shows mechanism of action of hormonal therapies. LHRH agonists or antagonists act at the level of hypothalamus thereby reducing the primary stimulus to the testis. Surgical removal of testis (Castration or bilateral orchidectomy) removes the primary source of testosterone hormone. First generation androgen blockers prevented binding of testosterone to its receptors. Newer androgen receptor blockers such as abiraterone and enzalutamide, in addition to receptor blockade, also prevent nuclear translocation of receptor-hormone complex.

22.4 Prostate Cancer Terminology [3]

22.4.1 Metastatic Hormone/Castration-Sensitive Prostate Cancer (mHSPC)

In this stage, cancer cells from the prostate have extended to the other parts of the body however it is still responsive to hormonal therapy.

22.4.2 Metastatic Hormone/Castration-Resistant Prostate Cancer (mCRPC)

It is metastatic prostate cancer that has developed resistance to hormonal therapy. Cancer cells continue to grow even in an androgen deprivation state (low testosterone).

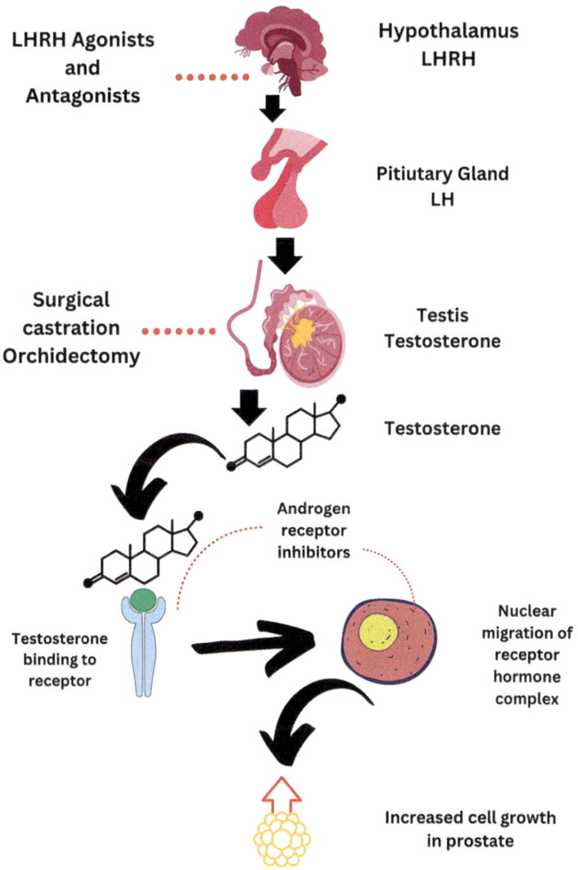

Fig. 22.1 Pituitary hypothalamus axis. (Source: Personal)

22.5 Indications for Hormonal Therapy

22.5.1 Localized (Early) Prostate Cancer

If cancer cells are limited to the anatomical boundaries of the prostate (called localized prostate cancer), hormone therapy is given along with the main treatment modality. It increases the effectiveness of the main therapy by shrinking the prostate size and the cancer inside it.

Duration of hormone therapy in such cases [3]**:**

- For 6 months along with radiotherapy for low-risk patients
- For high-risk localized prostate cancer patients for up to 3 years
- Before and after brachytherapy (a type of internal radiotherapy)
- Before focal therapy such as cryoablation or high-intensity focused ultrasound (HIFU) therapy

22.5.2 Locally Advanced Prostate Cancer

Hormone therapy may be given in patients with locally advanced (outside the prostate) or node-positive prostate cancers along with radiotherapy. Hormone therapy is usually offered for 6 months before radiotherapy and continued for a total of 3 years.

22.5.3 Advanced (Metastatic) Prostate Cancer

For advanced or metastatic prostate cancers, hormonal therapy may be needed lifelong. In this situation, hormone therapy can suppress the cancer for a few years but it can't be cured. However, it can help in the management of symptoms related to advanced cancer particularly bone pains. Duration of control varies with patients and depends on the inherent aggressiveness of the prostate cancer.

22.5.4 Recurrent Prostate Cancer

Prostate cancer that recurs after primary treatment (radiation or surgery) may also be treated with hormonal therapy.

22.6 Types of Hormonal Therapy

Broadly hormonal therapies can be classified on the basis of the mode of their administration.

1. Injection or implant
2. Oral medications
3. Surgical

22.6.1 Injections or Implants

These injections can be given in muscles (called intramuscular) or in the subcutaneous tissues (called subcutaneous injections). These act by blocking signals from the brain to the testicles to reduce the formation of testosterone. These injections are equally effective (compared to orchidectomy) for prostate cancer control. Using a needle a designated amount of drug is delivered below the skin or in the muscle. Depending on the type of drug, the injection may be given in the arm, abdomen (stomach area), thigh, or bottom (buttocks). Implants are tiny pellets of drugs that can be placed under the skin to slowly release the drug over time. Injection frequency varies according to the type of drug administered and it may vary between 1 to 6 months.

22.6.1.1 LHRH Agonists

Luteinizing hormone-releasing hormone (LHRH) agonists such as goserelin, leuprorelin acetate, triptorelin and buserelin acetate are the commonly used drugs for injections or implants.

After the first time injection, LHRH agonists can cause increased production of testosterone for a few days. During this temporary increase in testosterone levels, cancer cells can grow rapidly for a limited period, and cause symptoms to worsen—called the *flare phenomenon*. To counteract this flare phenomenon, a short duration of anti-androgen tablets (e.g. Bicalutamide) should be prescribed along with LHRH agonists.

22.6.1.2 GnRH Antagonists

Gonadotrophin-releasing hormone (GnRH) antagonists can be given subcutaneously every month or orally every day. Degarelix and Relugolix are the GnRH antagonists that are available and are usually given to patients with advanced prostate cancer. Unlike the agonists, antagonists don't cause an initial surge in testosterone. Antagonists don't require short duration of complementary anti-androgen tablets as they don't cause flare phenomenon. Starting dose for degarelix is 240 mg, which is given as two injections subcutaneously on two sites on the abdomen followed by a monthly injection of 80 mg. Recently, oral GnRH antagonist, i.e. Relugolix has been approved and it is started with a loading dose of 300 mg on day 1 followed by 120 mg every day.

22.6.2 Tablets

22.6.2.1 Androgen Synthesis Inhibitors or Antiandrogens

Anti-androgen tablets inhibit interaction of testosterone with the cancer cells by blocking the enzymes required for its production. They can be used alone, before, and, together with LHRH agonists. Antiandrogens include bicalutamide, cyproterone acetate and flutamide. These drugs when given alone are less effective, however, these are less prone to cause any sexual problems or bone thinning than other hormonal therapies.

22.6.2.2 Newer (Second-Generation) Hormone Therapy

These are the latest type of hormonal therapy called newer or second-generation hormonal therapy. They can be employed in both castration-resistant and castration-sensitive patients with metastatic prostate cancer. These latest drugs include abiraterone, enzalutamide, apalutamide and darolutamide.

Abiraterone tablets were most commonly given to men with castration-resistant metastatic prostate cancer. But now it's indication has expanded as a first line treatment option for castration sensitive metastatic prostate cancer as well.

Enzalutamide tablet like abiraterone may be offered to men with castration-resistant or castration-sensitive metastatic prostate cancer.

Darolutamide is offered to patients with non-metastatic castration resistant prostate cancer and in combination with Docetaxel (a type of chemotherapy) for metastatic hormone-sensitive prostate cancer.

Apalutamide tablets may be offered to some men with non-metastatic castration resistant prostate cancer and metastatic hormone-sensitive prostate cancer.

22.6.3 Surgery to Remove the Testicles (Orchidectomy)

Surgical removal of both the testis is called bilateral orchidectomy (or orchiectomy). Surgery is the quickest modality to reduce the testosterone levels form the body. Testosterone levels should drop to their lowest level in less than 12 hours after the surgery. Surgical removal is a single-time irreversible procedure and spares the need for repeated hormonal injections. As surgery is an irreversible procedure it's only offered to men who require lifelong hormonal therapy such as metastatic prostate cancers. Surgical procedure is generally well tolerated, however some minor short-term side effects including swelling and bruising of the scrotum can be seen.

22.7 Side Effects of Hormonal Therapy [4, 5]

Because androgens affect many other organs besides the prostate, it can have a wide range of side effects, including:

- Loss of libido (sexual desire)
- Loss of erectile function
- Hot flashes/warmth
- Reduced bone density and propensity to fractures
- Reduced muscle mass and strength
- Dyslipidemia (alteration in blood fats)
- Weight gain or insulin resistance (or increased insulin demand)
- Changes in the mood and fatigue

- Enlargement of breast tissue (called gynecomastia)

Apart from the generalized side effects noted above. Some drugs can cause certain side effects such as antiandrogens can cause loose motions, breast pain and nausea/vomiting. The antiandrogen drug flutamide may damage to the liver (called hepatotoxicity). Enzalutamide and apalutamide, can cause some neurological problems such as seizures (fits) and falls.

22.8 How to Mitigate Side Effects of Hormonal Therapy [3–5]?

The primary cause for majority of these side effects such as sexual and emotional are low levels of testosterone. These side effects would eventually go away if these drugs are stopped due to recovery in the levels of testosterone. However, sometimes, testosterone levels may not fully recover particularly in older men and men who received hormonal therapy for a long time. Therefore, in these patients side effects of hormonal therapy may not completely disappear.

Hormonal therapy induces certain physical changes in men such as loss of muscle and bone mass. Bone loss due to hormonal therapy may persist even after stoppage of these drugs. Therefore, men who are given hormonal therapy for a long period of time may be given medicines that slow down or reverse this bone loss. These drugs include bisphosphonates and denosumab that can be used to increase bone mineral density in men who are undergoing hormonal therapy. However, these drugs can cause a rare but serious side effect called osteonecrosis of the jaw.

Regular exercise, healthy diet, avoiding smoking and alcohol may help in reducing some of the side effects of the hormonal therapy such as muscle loss, weight gain, fatigue, and insulin resistance.

The sexual side effects, i.e. loss of libido and erection can be very difficult to mitigate for patients with prostate cancer on hormonal therapy. Some amount of recovery can be expected in these side effects after stopping the hormonal therapy with recovery of testosterone levels.

References

1. Gomella LG, Singh J, Lallas C, Trabulsi EJ. Hormone therapy in the management of prostate cancer: evidence-based approaches. Ther Adv Urol. 2010;2:171–81. Epub 2011/7/27.
2. Desai K, McManus JM, Sharifi N. Hormonal therapy for prostate cancer. Endocr Rev. 2021;42:354–73. Epub 2021/1/23.
3. Cornford P, van den Bergh RCN, Briers E, Van den Broeck T, Cumberbatch MG, De Santis M, et al. EAU-EANM-ESTRO-ESUR-SIOG guidelines on prostate cancer. Part II-2020 update: treatment of relapsing and metastatic prostate cancer. Eur Urol. 2021;79:263–82. Epub 2020/10/12.

4. Ziolkowska E, Zarzycka M, Wisniewski T, Zyromska A. The side effects of hormonal therapy at the patients with prostate cancer. Contemp Oncol (Pozn). 2012;16:491–7. Epub 2012/1/1.
5. Kumar RJ, Barqawi A, Crawford ED. Adverse events associated with hormonal therapy for prostate cancer. Rev Urol. 2005;7(Suppl 5):S37–43. Epub 2006/9/21.

Chapter 23
Chemotherapy in Prostate Cancer

Vineet Talwar and Arpit Jain

Chemotherapy prolongs life in metastatic Disease

23.1 How Chemotherapy Works?

- Chemotherapy drugs are usually given intravenously, but they can also be given orally or topically. The drugs travel through the bloodstream to reach cancer cells throughout the body. Chemotherapy drugs can affect both cancer cells and normal cells. However, cancer cells are more susceptible to the drugs because they divide and grow more rapidly than normal cells.
- The drugs used in chemotherapy are designed to destroy cancer cells by damaging their DNA or preventing them from dividing. Cancer cells divide more frequently than normal cells, so chemotherapy drugs are designed to target rapidly dividing cells. The drugs also target the blood vessels that supply nutrients to the cancer cells. By blocking these blood vessels, the drugs can starve the cancer cells of the nutrients they need to grow and survive. Unfortunately, chemotherapy drugs can also affect healthy cells that divide frequently, such as those in the hair, mouth, and digestive system.

V. Talwar (✉)
Director and Head, Department of Medical Oncology, Rajiv Gandhi Cancer Institute, New Delhi, Delhi, India

A. Jain
Consultant (Medical Oncology), Rajiv Gandhi Cancer Institute and Research Centre, Delhi, India

N. Gupta (ed.), *A Clinical Guide for Cancer Prostate Patient*,
https://doi.org/10.1007/978-981-97-5357-4_23

23.2 Benefits of Chemotherapy in Prostate Cancer Treatment

- Chemotherapy can be an effective treatment option for prostate cancer in several situations. For example, chemotherapy may be used:
  - In advanced prostate cancer: Chemotherapy can be used to shrink tumours and slow the progression of the cancer.
  - In recurrent prostate cancer: Chemotherapy can be given after other treatments have failed to control the cancer.
  - In combination with other treatments: Chemotherapy can be given in combination with other treatments such as surgery, radiation therapy, or hormone therapy to increase their effectiveness.
  - To relieve symptoms: Chemotherapy can be used to relieve pain and other symptoms caused by advanced prostate cancer.

23.3 Factors Affecting the Decision to Undergo Chemotherapy

- The decision to undergo chemotherapy is usually made by the patient and their healthcare team. Several factors can influence this decision, including:
  - The stage and aggressiveness of the cancer: Chemotherapy may be recommended for aggressive or advanced prostate cancer.
  - Other health conditions: Chemotherapy can be hard on the body, so other health conditions may need to be considered before starting treatment.
  - Age and overall health: Chemotherapy can be more difficult for older adults or those with other health conditions.
  - Personal preferences: Some patients may prefer to avoid chemotherapy because of the side effects or the impact on their quality of life.

23.4 Chemotherapy as Part of a Comprehensive Treatment Plan

- Chemotherapy is often used as part of a comprehensive treatment plan for prostate cancer. This may include a combination of surgery, radiation therapy, hormone therapy, and/or immunotherapy. The goal of a comprehensive treatment plan is to control the cancer and improve the patient's quality of life.
- In the past, chemotherapy was reserved for males whose advanced prostate cancer was no longer responding to ADT (androgen deprivation therapy). Chemotherapy is now often recommended in combination with ADT as the ini-

tial treatment for males whose cancer has already spread outside of the prostate, typically to the bones or other organs.

- Common chemotherapy options include:
  - Docetaxel
  - Cabazitaxel
  - Cisplatin
  - Carboplatin
  - Mitoxantrone
- These drugs have been shown to improve overall survival rates in patients with metastatic castrate-resistant prostate cancer (mCRPC). This chapter will review the current state of chemotherapy in prostate cancer treatment based on evidence from clinical trials and clinical practice.

23.4.1 Chemotherapy in Advanced Prostate Cancer

- Chemotherapy can be an effective treatment option for advanced prostate cancer. In this situation, chemotherapy is used to shrink tumours and slow the progression of the cancer. Chemotherapy can also be used to relieve symptoms such as pain and other discomfort.

23.4.2 Chemotherapy in Recurrent Prostate Cancer

- Chemotherapy can also be used to treat recurrent prostate cancer. Recurrent prostate cancer is cancer that has come back after initial treatment by surgery or Radiotherapy. Chemotherapy can be given after other treatments like hormone therapy has failed to control the cancer.

23.4.3 Docetaxel

- Docetaxel is a chemotherapy medicine used along with hormone therapy in patients with advanced prostate cancer. Docetaxel is an option for some who are taking ADT for the first time. Docetaxel is also used to treat metastases after ADT fails to stop cancer growth. Though docetaxel can't cure prostate cancer, it can help people live longer as well as reduce their pain and other symptoms.
- The TAX327 trial compared docetaxel, a taxane-based chemotherapy drug, to mitoxantrone in patients with advanced prostate cancer. The study found that docetaxel improved overall survival and quality of life compared to mitoxantrone [1].

- Docetaxel is one of the most commonly used drug for cancer prostate. Usually, it is well tolerated by the patients. According to the patient age, kidney function and health conditions, dose, frequency and cycles of therapy is given under medical supervision.

23.4.4 *Cabazitaxel*

- Cabazitaxel is one such drug that has been approved for use in advanced prostate cancer patients who have already received chemotherapy with docetaxel. Cabazitaxel works by binding to microtubules in the cell, which are essential for cell division and growth. This inhibits cell replication and ultimately leads to their death [2].
- Clinical trials have shown that cabazitaxel can improve survival rates and quality of life in advanced prostate cancer patients when combined with prednisone—a steroid medication commonly used alongside chemotherapy.
- A follow-up study, called TROPIC, evaluated the effectiveness of another taxane-based chemotherapy drug, cabazitaxel, in patients with advanced prostate cancer who had already received docetaxel. The study found that cabazitaxel improved overall survival compared to mitoxantrone.

23.4.5 *Mitoxantrone*

- Mitoxantrone is used to relieve pain and decrease the need for pain medications. It's an option for men who aren't able to tolerate other therapies.

23.4.6 *Cisplatin and Carboplatin*

- Cisplatin and carboplatin are chemotherapy drugs made from platinum. These are sometimes used for patients with very advanced or aggressive cancer. Usually, either cisplatin or carboplatin is combined with another chemotherapy medicine such as cabazitaxel or docetaxel [3].

23.5 Alternatives to Chemotherapy in Prostate Cancer Treatment

- Chemotherapy is not the only treatment option for prostate cancer. Other treatment options include:
  - Surgery: Surgery may be recommended to remove the cancerous prostate gland.
  - Radiation therapy: Radiation therapy uses high-energy beams to kill cancer cells.
  - Hormone therapy: Hormone therapy can be used to block the production of hormones that fuel the growth of prostate cancer cells.
  - Targeted therapy: Certain genes, such as BRCA1, BRCA2, and several others, are linked with DNA-repair gene defects, and the presence of any of these can allow treatment with Olaparib/Rucaparib [4].
  - Immunotherapy: Immunotherapy uses the body's immune system to fight cancer cells.

23.6 Risks and Side Effects of Chemotherapy

- Although chemotherapy can be an effective treatment option, it can also have significant side effects.
- The most common side effects of chemotherapy are: Nausea and vomiting: Chemotherapy drugs can cause nausea and vomiting, which can be managed with anti-nausea medications.
  - Hair loss: Chemotherapy drugs can cause hair loss on the scalp and other parts of the body.
  - Fatigue: Chemotherapy can cause fatigue, which can be managed with rest and exercise.
  - Infection: Chemotherapy can weaken the immune system, making it more susceptible to infection.
  - Anemia: Chemotherapy can cause a decrease in red blood cells, which can cause fatigue and shortness of breath.
  - Neuropathy: Chemotherapy can cause nerve damage, which can cause numbness, tingling, or pain in the hands and feet.

23.7 Conclusion

- In conclusion, chemotherapy can be an effective treatment option for prostate cancer, but it also comes with risks and side effects that should be carefully considered. Patients should work with their healthcare team to determine the best treatment plan based on their individual needs and goals.
- Clinical trials and clinical practice have shown that chemotherapy can improve overall survival in patients with advanced or metastatic prostate cancer. While side effects can occur, they can be managed through careful monitoring and management. As research into prostate cancer continues, it is likely that new and more effective chemotherapy drugs will be developed that can further improve the outcomes for patients with this disease.
- Alternatives to chemotherapy such as surgery, radiation therapy, and hormone therapy may be more appropriate for some patients depending on the stage and grade of the cancer.
- Regardless of the treatment plan chosen, patients should prioritize their quality of life and work closely with their healthcare team to manage any side effects and complications that may arise.

References

1. Tannock IF, de Wit R, Berry WR, Horti J, Pluzanska A, Chi KN, Oudard S, Théodore C, James ND, Turesson I, Rosenthal MA, Eisenberger MA, TAX 327 Investigators. Docetaxel plus prednisone or mitoxantrone plus prednisone for advanced prostate cancer. N Engl J Med. 2004;351(15):1502–12. https://doi.org/10.1056/NEJMoa040720.
2. de Bono JS, Oudard S, Ozguroglu M, Hansen S, Machiels JP, Kocak I, Gravis G, Bodrogi I, Mackenzie MJ, Shen L, Roessner M, Gupta S, Sartor AO, TROPIC Investigators. Prednisone plus cabazitaxel or mitoxantrone for metastatic castration-resistant prostate cancer progressing after docetaxel treatment: a randomised open-label trial. Lancet. 2010;376(9747):1147–54. https://doi.org/10.1016/S0140-6736(10)61389-X.
3. Hager S, Ackermann CJ, Joerger M, Gillessen S, Omlin A. Anti-tumour activity of platinum compounds in advanced prostate cancer—a systematic literature review. Ann Oncol. 2016;27(6):975–84. https://doi.org/10.1093/annonc/mdw156. Epub 2016 Apr 6.
4. Hussain M, Mateo J, Fizazi K, Saad F, Shore N, Sandhu S, Chi KN, Sartor O, Agarwal N, Olmos D, Thiery-Vuillemin A, Twardowski P, Roubaud G, Özgüroğlu M, Kang J, Burgents J, Gresty C, Corcoran C, Adelman CA, de Bono J, PROfound Trial Investigators. Survival with olaparib in metastatic castration-resistant prostate cancer. N Engl J Med. 2020;383(24):2345–57. https://doi.org/10.1056/NEJMoa2022485. Epub 2020 Sep 20.

Chapter 24
Immunotherapy in Prostate Cancer

Ashok Vaid and Ankur Mudgal

Immunotherapy is an emerging field in management of cancer prostate

24.1 What Is Immune System?

The immune system protects the human body from various threats like microorganisms (bacteria, viruses, fungi) and cancer cells.

There are mainly two ways through which immune system protects human body:

- Physical barriers: Skin, cornea, lining of the gastrointestinal tract, urinary bladder and reproductive tracts.
- Organs producing specified immune cells: Lymphatic system, bone marrow, spleen and thymus gland.

Physical barriers acts as first line of defense and white blood cells forms next line of defense which fights microorganisms and cancer cells. White blood cells (also known as leuckocytes) are specialized immune cells produced by bone marrow and thymus gland.

A. Vaid (✉) · A. Mudgal
Medical Oncology and Hematology, Medanta—The Medicity, Gurugram, India

N. Gupta (ed.), *A Clinical Guide for Cancer Prostate Patient*,
https://doi.org/10.1007/978-981-97-5357-4_24

24.2 What Are Different Types of Immune Response in Human Body?

There are mainly two types of immune responses:

- Innate immune response: It is a non-specific quick response when the body encounters a harmful pathogen.
- Acquired immune response: It is a more specific response which the body slowly learns and acquires when it encounters some abnormal cell and reacts more strongly next time it encounters with the same cell. B cells and T cells plays important role in this type of immune response.

24.3 How Do Immune System Kill Cancer Cells?

- The immune system continuously keeps check on abnormal cell growth and destroy those cells without being a person aware about it.
- Dendritic cells are white blood cells that works as antigen presenting cells (APC). These cells are responsible for both waste removal and alerting other immune system cells whenever body encounters a foreign antigen.
- T cells are white blood cells that are activated by APCs by presenting tumor antigens to T cells. They recognize tumor antigen as foreign antigen and destroys the cell.
- APC's stimulate T cells via a "lock and key" mechanism facilitating them to distinguish, attack and kill tumor cells.
- When they reach the tumor, they can recognize it, and destroy it by releasing granzymes or perforin, which can rupture the cell membrane.

24.4 What Are Immune Check Points?

- The activation of immune cells is tightly regulated by both stimulatory and inhibitory molecules on both APC and T cells, referred to as "immune checkpoint" molecules because they act as a physiologic brake on T cells.
- Cytotoxic T-lymphocyte-associated antigen 4 (CTLA 4) and programmed cell death-1 (PD-1) are examples of co-inhibitory or "immune checkpoint" molecules.
- These molecules are present to control immune response and prevent autoimmunity (over activation of immune system) but in context to cancer these molecules results in immune tolerance.

24.5 What Is the Concept Behind Immunotherapy?

- While the growth and division of tumor cells are directly impacted by chemotherapy or tumor-targeted medications, but immunotherapy take advantage of the body's own immune system's capacity to fight and eradicate tumor cells.
- Immunotherapy can be divided into two main categories:
  - Active immunity: In this immune cells are trained to recognize and act against tumor cells, e.g. Cancer vaccine (Sipuleucel T)
  - Passive immunity: In this body's own immune response is enhanced, e.g. Immune checkpoint inhibitors (ICI).
- Checkpoints are intended to stop the immune response in order to avoid autoimmunity and injury to normal cells, but tumor cells circumvents these mechanisms by "inhibiting" T cells after they have recognized the tumor, limiting attack and cell-killing activity.
- The body's anti-tumor immune response is boosted by the use of CTLA-4 inhibitors, PD-1 pathway inhibitors, or PD-L1 (Program cell death ligand-1) inhibitors, which stop this deactivation (Table 24.1).

24.6 How Does Immunotherapy Differ from Chemotherapy and Tumor-Targeted Drugs?

- Chemotherapy destroys rapidly dividing cells in a non-selective manner and act on both cancer cells and normal cells which causes more side effects and damage to hair follicles, bone marrow and mucosa of gastrointestinal tract.
- Targeted agents are designed to specifically act on molecular targets in tumor cells. These drugs are used in tumors over expressing those targets and expected to have less side effects.
- Immunotherapy works by activating the body's immune system to kill cancer cells. This sometimes leads to over-activation of our immune system and can lead to autoimmune side effects.

Table 24.1 Mechanism of action for immunotherapy drugs

Mechanism of action	Drugs
Anti CTLA-4 inhibitors	Ipilimumab
Anti PD 1 inhibitors	Pembrolizumab
	Nivolumab
Anti PDL 1 inhibitors	Atezolizumab
	Avelumab
	Durvalumab

24.7 What Are the Indications of Immunotherapy in Prostate Cancer?

- There is very limited use of immunotherapy in prostate cancer which is only limited to metastatic castration resistant prostate cancer (Advanced prostate cancer that has become resistant to first line hormone therapy).

24.8 What Are Various Biomarkers for Immunotherapy?

- A biomarker is used to get an insight into genetic makeup of tumor cells, their behavior, and interactions with immune system. They help in predicting response to immunotherapy agents and selecting patients who are more likely to responds to these drugs.
- PD-1/PD-L1 expression: It is a test used to assess the level of PD-L1 protein expression on tumor cells or immune cells inside a tumor. It is present on cell surfaces and controls the immune system's response against tumor cells. It is measured on tumor tissue with the help of immunohistochemistry (IHC). Through it's over expression tumor cells evade immune response by inhibiting T cells. Tumor cells with increased expression of these molecules are more likely to respond to immunotherapy. It is still not an approved test for use of immunotherapy in patients with prostate cancer.
- Tumor Mutation Burden (TMB): It is the measure of total number of mutations that a tumor has accumulated. A tumor's tendency to express mutant proteins and neoantigens, which the immune system can then target, increases with the number of mutations it exhibits. Patients are given consideration for immunotherapy if they have ten or more somatic mutations per megabase of the probed genomic sequence.
- Microsatellite instability (MSI)/Deficient mismatch repair (dMMR): Genes that control cell proliferation and differentiation undergo alteration in the development of cancer. These genetic changes include, for instance, altering the DNA sequence of a gene (referred to as a mutation), altering the number of chromosomes or breaking them (referred to as chromosomal instability), and altering the length of particular repeat sequences in the DNA (referred to as microsatellite instability). These mutations result in the production of mutant proteins, and mutation-derived antigens that can be recognized by CD8+ T cells and also by the immune system. Anti-PD-1 immunotherapy, independent of the type of malignancy, became the sole therapy approved for patients with MSI-high/dMMR solid tumors in May 2017.

Unfortunately, dMMR/MSI-H is relatively uncommon in prostate cancer (ranging from <1% to 8%). As example one database study of 741 patients with metastatic CRPC, all of whom underwent comprehensive genomic profiling revealed 2 with

MSI-H/low TMB disease (<1%) and 20 (2.7%) with MSI-H and TMB ≥10 mutations per megabase; an additional 24 (3.2%) had microsatellite stable tumors but a TMB ≥10 mutations per megabase [1, 2].

24.9 What Are the Various Types of Immunotherapy Available in Prostate Cancer?

- Immunotherapy drugs approved for use in prostate cancer are Sipuleucel-T, Pembrolizumab and Dostarlimab.

24.10 What is Sipuleucel-T and How It Works?

- It is the first autologous cellular therapeutic vaccine approved for any tumor type in 2010.
- It acts by boosting patient's immune cells against prostate cancer cells.
- It is prepared in three steps:
  - First step: A procedure known as leukapheresis is used to collect the patient's own immune cells, primarily APCs. Than they are isolated and collected using a specialized machine while withdrawing blood from the patient.
  - Second step: The gathered APCs are transported to a lab, where they go through an activation and modification process. Prostatic acid phosphatase-granulocyte-macrophage colony-stimulating factor, or PAP-GM-CSF, is a fusion protein that combines a protein present in prostate cancer cells (PAP) with an immune-stimulating factor (GM-CSF). The APCs are stimulated by this exposure, which improves their ability to identify and target prostate cancer cells.
  - Third step: These activated immune cells are then reinfused in patients body where they act against prostate cancer cells.

24.11 When Do We Use Sipuleucel-T in Prostate Cancer?

- It is reserved for patients with asymptomatic or minimally symptomatic slowly progressive metastatic CRPC who do not have visceral (especially hepatic) metastasis or cancer pain requiring opioids, but even in this population, other treatments may also be appropriate.
- It is given in three doses with around 1–2 weeks gap. Each dose containing more than 50 million activated cells.

- Patient is pre-medicated prior to infusion of each vaccine to decrease chances of any allergic reaction. Each intravenous infusion is completed in around 60 min and patient is kept for monitoring for about 30 min. Again the same process is repeated in 1–2 weeks time.
- Efficacy: Initial placebo controlled trials showed 4-month improvement in median overall survival (OS) in males with asymptomatic or minimally symptomatic metastatic disease [3, 4].
- Side effects: Chills, lethargy, fever, back discomfort, nausea, joint soreness, and headache were the most prevalent side effects (>15%). These effects are usually transient and resolve within couple of days. Some of the very rare adverse effects are rhabdomyolysis, myositis and severe infusion reaction.

24.12 Where Do PD-1 Pathway Inhibitors Are Used in Prostate Cancer and How They Work?

- Mainly two anti PD1 inhibitors are currently approved in prostate cancer; Pembrolizumab and Dostarlimab. These immunotherapy drugs are monoclonal antibodies. They are used in patients of prostate cancer with specific genetic mutations and works by boosting body's immune cells capacity to identify and kill cancer cells.
- The US Food and Drug Administration (FDA) gave approval to these drugs for treatment of refractory solid tumors (including prostate cancer) with no other satisfactory treatment alternatives that have dMMR. Trial which lead to approval had only two patients with prostate cancer [5].
- T cells are inactivated through PD-1 receptors present on its surface after interaction with PD-L1 ligand present on tumor cells. This T cell inactivation acts as brake on immune cells mediated tumor eradication.
- Pembrolizumab is an option for treatment of patients with metastatic CRPC who have tumors with dMMR or high levels of TMB (≥10 Mut/Mb) that has progressed following prior treatment and for whom there are no satisfactory alternative treatment options.
- In a retrospective analysis of 741 patients with metastatic CRPC patients with high TMB (5% of patients) showed better survival with immunotherapy in comparison with chemotherapy [6, 7].
- Patients planned for immunotherapy should not have any active autoimmune disease that cannot be addressed with hormone replacement therapy.
- Pembrolizumab is administered intravenously either as 200 mg every 3 weeks or 400 mg every 6 weeks. It is diluted in either normal saline and infused over 30 min.
- Dostarlimab is administered as 500 mg every 3 weeks through four cycles then 1000 mg every 6 weeks.

24.13 What Are the Side Effects Related to PD-1 Inhibitors?

- Immunotherapy drugs mostly cause side effects related to skin, intestines, lungs, liver and other endocrine organs (such as the pituitary gland or thyroid gland).
- Most of the side effects are mild to moderate and reversible if detected and treated early.
- Fatigue: Many patients on immunotherapy experiences fatigue and it's important to rule out other causes like thyroid or pituitary disorders.
- Skin reactions which include extensive rash or itching.
- Gastrointestinal complications like diarrhea associated with abdominal pain or stools containing blood or mucus.
- Endocrine dysfunction presents in form of weight gain, fatigue, nausea, vomiting, excessive thirst and excessive urination.
- Respiratory complaints include cough, shortness of breath, chest pain.
- Other less common complaints are headache, state of confusion, muscle soreness, pain and swelling in joints, unexplained fever and blurred vision.

24.14 When Are These Side Effects Most Likely to Appear and How Common Are They?

- These side effects usually appear within weeks to 3 months of starting treatment.
- It is seen that the timeline for immunotherapy related adverse events corresponds the development of the body's immune response to tumor as it is enhanced by PD-1 inhibitors. Over activation of this response that leads to autoimmunity.

24.15 What Are Other Immunotherapeutic Approaches Under Evaluation in Prostate Cancer?

- A CTLA-4 inhibitor (Ipilimumab) had shown decrease in prostate specific antigen (PSA) levels and objective response in patients with CRPC. CTLA 4 is also an immune checkpoint other than PDL-1 and it also acts by inhibiting signals to T cells. However, two large phase 3 trials failed to show any survival benefit [8].
- In roughly 1300 males with metastatic CRPC, PROSTVAC, an investigational therapeutic pox-viral vaccination, was tried. Despite research showing an increased immune response in the peripheral blood and immune infiltrates within the tumor, a randomized controlled investigation found no evidence that the vaccine increased survival [9].
- Chimeric antigen receptor T (CAR-T) cells and bispecific antibodies are under early trials but without much success [8].

- Given the inadequate immune response to prostate cancer in vivo, it is not surprising that immunotherapeutic approaches (such as immune checkpoint inhibitors [ICIs] and CAR-T cells) that have been so successfully used to treat other solid tumors and hematologic malignancies have not demonstrated comparable clinical efficacy in unselected prostate cancers.

24.16 Conclusion

In the treatment of prostate cancer, immunotherapy has become a ground-breaking strategy in a very selected group of patients. Innovative approaches for treatment of this illness include immune checkpoint inhibitors, therapeutic cancer vaccines, and other methods. An important milestone was reached with the approval of pembrolizumab for a subset of patients with advanced prostate cancer, and studies continue to identify novel treatment approaches. As the field develops, the incorporation of immunotherapy with other therapeutic modalities offers enormous potential for enhancing outcomes and increasing the quality of life for patients with prostate cancer.

References

1. Rodrigues DN, Rescigno P, Liu D, Yuan W, Carreira S, Lambros MB, Seed G, Mateo J, Riisnaes R, Mullane S, Margolis C. Immunogenomic analyses associate immunological alterations with mismatch repair defects in prostate cancer. J Clin Invest. 2018;128(10):4441–53.
2. Mahoney KM, Atkins MB. Prognostic and predictive markers for the new immunotherapies. Oncology (Williston Park, NY). 2014;28:39–48.
3. Madan RA, Antonarakis ES, Drake CG, Fong L, Yu EY, McNeel DG, Lin DW, Chang NN, Sheikh NA, Gulley JL. Putting the pieces together: completing the mechanism of action jigsaw for sipuleucel-T. J Natl Cancer Inst. 2020;112(6):562–73.
4. Kantoff PW, Higano CS, Shore ND, Berger ER, Small EJ, Penson DF, Redfern CH, Ferrari AC, Dreicer R, Sims RB, Xu Y. Sipuleucel-T immunotherapy for castration-resistant prostate cancer. N Engl J Med. 2010;363(5):411–22.
5. Lemery S, Keegan P, Pazdur R. First FDA approval agnostic of cancer site-when a biomarker defines the indication. N Engl J Med. 2017;377(15):1409–12.
6. Graf RP, Fisher V, Weberpals J, Gjoerup O, Tierno MB, Huang RS, Sayegh N, Lin DI, Raskina K, Schrock AB, Severson E. Comparative effectiveness of immune checkpoint inhibitors vs chemotherapy by tumor mutational burden in metastatic castration-resistant prostate cancer. JAMA Netw Open. 2022;5(3):e225394.
7. Hansen A, Massard C, Ott P, Haas N, Lopez J, Ejadi S, Wallmark J, Keam B, Delord JP, Aggarwal R, et al. Pembrolizumab for patients with advanced prostate adenocarcinoma: preliminary results from the KEYNOTE-028 study. Ann Oncol. 2016;27:vi243.
8. Bilusic M, Madan RA, Gulley JL. Immunotherapy of prostate cancer: facts and hopes prostate cancer immunotherapy. Clin Cancer Res. 2017;23(22):6764–70.
9. Gulley JL, Madan RA, Tsang KY, Jochems C, Marté JL, Farsaci B, Tucker JA, Hodge JW, Liewehr DJ, Steinberg SM, Heery CR. Immune impact induced by PROSTVAC (PSA-TRICOM), a therapeutic vaccine for prostate cancer. Cancer Immunol Res. 2014;2(2):133–41.

Chapter 25
Radiopharmaceuticals for Prostate Cancer Management

Manas Sahoo and Lokesh Kumar

Lutetium based therapy is emerging for the treatment of metastatic disease (NPG)

25.1 Introduction

- Usage of various radioactive material referred as radiopharmaceutical for diagnosing and treating multiple diseases have become very popular in the recent times. These radiopharmaceuticals are administered inside the body through different routes like intravenous, oral etc. Some time is allowed till the radiopharmaceutical reaches the target site.

25.1.1 Application of Radiopharmaceutical in Prostate Cancer

- Radiopharmaceuticals are used to diagnose and stage prostate cancer
- Radiopharmaceuticals are used to treat prostate cancer and for bone pain palliation in advanced skeletal metastatic disease

25.1.2 Application of Radiopharmaceutical in Diagnosis and Staging of Prostate Cancer

Commonly used scans for diagnosis and staging of prostate cancer is

1. ^{99m}TC (Technetium) MDP bone scan
2. ^{68}Ga (Gallium)-PSMA PET/CT scan or ^{18}F-PSMA PET/CT scan

M. Sahoo (✉) · L. Kumar
Nuclear Medicine, Medanta—The Medicity, Gurugram, Gurugram, India
e-mail: manas.sahoo@medanta.org

N. Gupta (ed.), *A Clinical Guide for Cancer Prostate Patient*,
https://doi.org/10.1007/978-981-97-5357-4_25

25.2 What Is ^{99m}TC MDP Bone Scan?

- ^{99m}TC MDP bone scan is a whole-body scan to evaluate the bone status, in which ^{99m}TC MDP is given intravenously to localise bone-forming (osteoblastic) skeletal metastases. With the help of a gamma camera, a specialised camera that detects gamma rays scans the patient, to detect metastases.

25.2.1 Role of ^{99m}TC MDP Bone Scan in Prostate Cancer

- Because bone metastases from prostate cancer are mostly osteoblastic and have a prevalence nearly up to 70% [1], they are easily spotted on bone scans
- According to NCCN guidelines, bone scan in prostate cancer is indicated for symptomatic patients and/or those with life expectancy >5 years when they have following [2]

  (a) T1 disease with serum PSA >20 ng/mL
  (b) T2 disease with serum PSA >10 ng/mL
  (c) Gleason score ≥8 or T3 or T4 disease or symptomatic disease [2]

25.2.2 Procedure for Bone Scan

- Patient should be well hydrated and after taking formal written consent, 20 mci of ^{99m}TC MDP is injected intravenously and whole body scintigraphy is performed after 3–4 h under gamma camera.

25.2.3 Pros and Cons of Bone Scan

Pros

- Can detect bony metastases at early stage
- High sensitivity
- High negative predictive value
- Easy to perform
- Widely available
- Less costly
- No fasting is required

Cons

- Time consuming
- Less sensitive for localising osteolytic (bone dissolving) lesions
- Less resolution, so poor in detection of lesion below 3 mm
- Radiation exposure (Figs. 25.1 and 25.2)

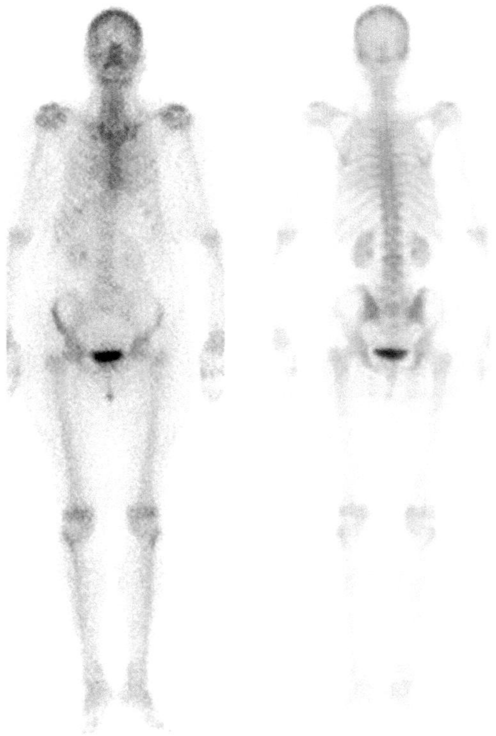

Fig. 25.1 Anterior and posterior image of normal bone scan in a 57 year prostate cancer patient

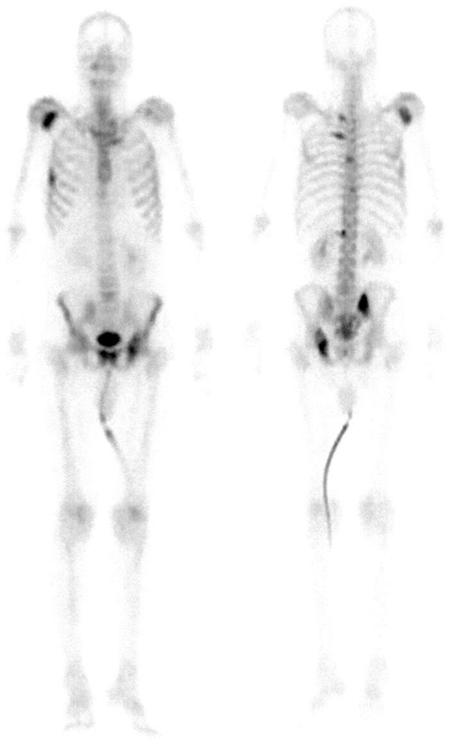

Fig. 25.2 Bone scan showing multiple skeletal metastases in a 63 year prostate cancer patient. (Source: Self)

25.3 What Is $^{68}Ga/^{18}F$ PSMA PET-CT Scan

- $^{68}Ga/^{18}F$ PSMA-ligand is a specific radiolabeled antibody. After the administration of $^{68}Ga/^{18}F$ PSMA-ligand, these antibodies binds specifically to PSMA (prostate specific membrane antigen) which is a type of transmembrane protein. Expression of this protein is significantly increased in prostate cancer and metastases of prostate cancer. So it is a type of antigen antibody reaction, after intravenous administration $^{68}Ga/^{18}F$ PSMA-ligand. This results in detection of the primary as well as secondary disease sites as the cancer cells overexpress PSMA.

25.3.1 How PET/CT Scanner Work

PET/CT scanner is a specialised machine which detects high energy photons. After administration of positron emitting radionuclide, they localise at various organs in body *depend upon type of radiopharmaceutical. After localisation, the positron emitted by these* positron emitting radiopharmaceutical combines with nearby electron in the body and loses its energy in the form of two high energy photons 180°

apart. These photons are detected by the specialized detector present in PET/CT scanner and PET image is formed after various electronic processes.

Simultaneously whole body CT scan is performed by CT scanner present in PET/CT machine and this CT scan image is fused with PET images.

25.3.2 Indication of PSMA PET-CT Scan in Prostate Cancer

According to appropriate use criteria PSMA PET-CT scan is indicated in [3]:

1. Newly diagnosed prostate cancer with unfavourable intermediate and high or very high risk
2. Newly diagnosed prostate cancer with unfavourable intermediate and high or very high risk along with negative/equivocal or oligometastatic disease on conventional imaging
3. Conventional imaging suggestive of widely spread metastatic disease in newly diagnosed case of prostate cancer
4. Persistent rising PSA or rising of PSA from undetectable serum PSA levels after radical prostatectomy
5. Rising of PSA from undetectable serum PSA levels after definitive radiotherapy
6. Non metastatic castrate resistant prostate carcinoma on conventional imaging
7. Post treatment rise of serum PSA metastatic castration resistant prostate carcinoma
8. Treatment response evaluation

25.3.3 Pros of PSMA PET-CT Scan [4, 5]

- Can guide the site of biopsy in previous negative biopsy in patient with high suspicion of prostate carcinoma
- PSMA PET-CT scan has high specificity and positive predictive value for lymph node and bony spread with respect to conventional imaging
- Can help in detecting the site of production of PSA after radical prostatectomy
- Can help in differentiating between recurrence and scar tissue
- Can help in stereotactic radiotherapy planning
- PSMA therapy can be given in PSMA expressing lesions in metastatic prostate castrate resistant prostate carcinoma
- Can detect both osteolytic and osteoblastic lesions.
- Because of better resolution of PET-CT, it can detect small lesions

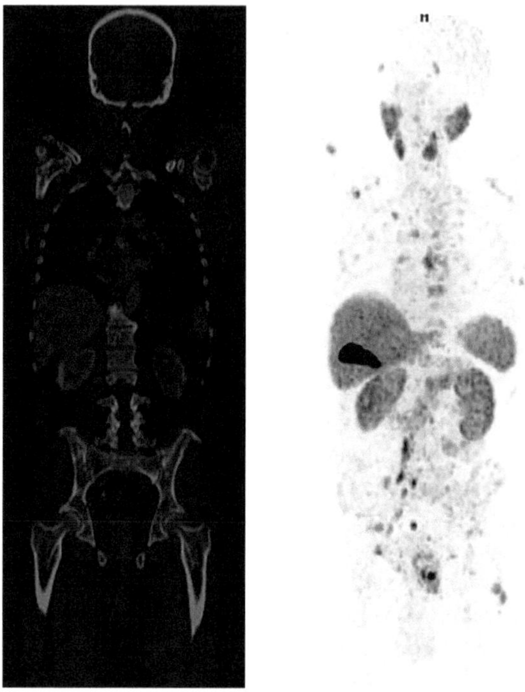

Fig. 25.3 GA-PSMA PET/CT in a 68 year patient of adenocarcinoma prostate showing PSMA expressing primary involving prostate with PSMA expressing skeletal and lymph nodal metastases. (Source: Self)

25.3.4 Cons of PSMA PET-CT Scan

- Radiation exposure but benefits are more than risk associated with radiation
- Few patient may develop allergic reaction to contrast used if contrast enhance CT scan is performed but in most of the cases these allergic reactions are minor and can be managed easily
- Very small lesions/non PSMA expressing lesions are not picked up (Fig. 25.3)

25.4 Role of Radionuclides in Treatment of Prostate Cancer

- Radionuclide can be used for bone pain palliation and treatment of metastatic castrate resistant prostate carcinoma

25.4.1 Bone Pain Palliation in Prostate Carcinoma [6]

- It is a palliative care in the patients suffering from terminal illness. It focuses on providing relief from pain and physical stress associated with serious illness in patients whose life expectancy >3 months [6] by using bone seeking radioactive elements/compounds

- In prostate cancer associated with bony metastasis, because of which patient has unbearable bone pain so bone pain palliation can improve quality of life in these patient

25.4.2 Indication of Bone Pain Palliation

- It is indicated in patient in patient with multifocal painful bony metastasis (osteoblastic or mixed osteoblastic/lytic) refractory to analgesics (pain killing medicine) and other treatment modalities [6]

25.4.3 Prerequisite for Bone Pain Palliation Radionuclide Therapy [6]

- Patients must have multiple painful bony metastases [6]
- Bony metastasis show intense uptake on bone scan performed not more than 4–8 weeks before the therapy [6]
- Patient must not have purely lytic bony lesion with impending risk of pathological fracture or spinal cord compression
- Adequate haematological parameters obtained no longer than 1–2 weeks before therapy
- Life expectancy must be at least more than 4–6 weeks preferably 3 months because bone palliation therapy takes around days to weeks to show its effect. So bone pain palliation therapy is considered in patients with relative longer life expectancy [3]

25.4.4 Radionuclides Used in Bone Pain Palliation [7]

- They are two types of radionuclides used in bone pain palliation depending upon types of radiation released by them

  1. Beta ray emitting
  2. Alpha ray emitting

     (a) Beta ray emitting radionuclides include Phosphorus-32 pyrophosphate, Strontium-89 chloride, Samarium-153 EDTMP, Lutetium-177 EDTMP, etc.
     (b) Alpha ray emitting radiopharmaceuticals in bone pain palliation mainly include Radium-223 chloride. Radium-223 chloride can be used painful osteoblastic/mixed Osteoblastic and osteolytic metastasis in castration resistant prostate carcinoma with no evidence of visceral metastases. Alpha therapy with Radium-223 chloride can also Increase the life expectancy of patient [7]

25.4.5 Pros of Bone Pain Palliation

1. Specific uptake of radiopharmaceuticals at the painful osteoblastic lesions in body
2. Rapid pain relief
3. Prolonged pain relief
4. Very high target to background ratio so less radiation exposure to surrounding organ

25.4.6 Cons of Bone Pain Palliation

1. Haematological toxicity is the main side effect but in most of the case it is temporary
2. Transient mild self-limiting increase in pain (flare phenomenon) but can be prevented by injectable steroid
3. Renal toxicity can be prevented by dose of appropriate radiopharmaceutical and good hydration

25.4.7 Contraindication of Bone Pain Palliation in Prostate Cancer [6]

Patient is evaluated carefully before treating to asses benefit vs risk

- Deranged renal profile—since these radio-labelled bone seeking agents mainly excreted through kidneys in the urine so if renal function is deranged probability of bone marrow toxicity will increase. So patients with GFR <30 mL/min and Serum creatinine >2 mg/dL are unfit for bone pain palliation therapy.
- Deranged haematological parameters:

  1. Haemoglobin <9 g/dL
  2. Platelets <100,000/mm^3
  3. White blood cell count <3500/mm^3

- Impending spinal cord compression and impending vertebral collapse is a relative contraindication.
- **Super scan**—It is a type of a bone scan in which intense uptake noted skeletal system and no significant uptake in soft tissue like kidney so it will increase the risk of the bone marrow toxicity and bone pain palliation therapy is an absolute contraindicated in these patient.

Note—bone marrow metastases is not a contraindication for bone pain palliation therapy.

25.4.8 *Patient Preparation [6]*

No specific patient preparation is required

- Informed written consent is taken informing that
    1. It is only a palliative care and 50–90% of patient benefited by the therapy.
    2. Pain can flare up for initial few days and pain reduction most likely to occur after 4 weeks of therapy.
    3. Analgesic effect of bone pain palliation therapy generally last for 2–6 months.
- Proper hydration is maintained before therapy

25.4.9 *Radiopharmaceutical Administration [6]*

- Appropriate radiopharmaceutical is administered slowly via cannula followed by flushing of cannula via 0.9% normal saline
- Doses of various radio-pharmaceuticals used in bone-pain palliation are:
    1. Strontium-89: 1.5–2.2 MBq/kg of body weight
    2. Samarium-153: 37 MBq/kg
    3. Phosphorus-32: 185–370 MBq
- Prophylactic corticosteroid injection is considered in patients with cervico-dorsal spinal metastases with impending risk of spinal cord compression

25.4.10 *Side Effect of Bone Pain Palliation Therapy [6]*

- Temporary flare up of pain can be seen in 10% of patient in initial 72 h of treatment but can be managed easily by analgesics and flaring up of pain is an indicator of good clinical response
- Decrease platelets and white blood cells are noticed and recovery can occur in between 3–12 weeks depending upon used radiopharmaceutical
- Flushing sensation can be noticed case strontium-89 based bone palliation therapy and can be avoided by slow IV infusion

25.4.11 *Radiation Safety [6]*

- Patient must be appropriately hydrated before and after the therapy
- Radiopharmaceutical is excreted in urine during first 2–3 days administration of radio pharmaceutical, so patient should avoid soiling of clothes if contaminated clothes should washed separately

- Double toilet flush is recommended
- If contamination of hand occurs, hand should be washed with cold water without scrubbing of hand
- If patient have urine incontinence, catheterisation is done before the therapy

25.5 Radionuclide Therapy in Treatment of Prostate Carcinoma

- Radionuclide therapy is indicated in metastatic castrate resistant prostate carcinoma or metastatic prostate carcinoma which is not responding to conventional first line therapy, i.e. Luteinizing hormone releasing analogues, anti-androgenic agents, chemotherapeutic agents

25.5.1 Radionuclide Therapy in Castration Resistant Prostate Carcinoma [8]

- PSMA can be labelled with various beta rays emitting radionuclide ex Iodine-131, Yttrium-90, Copper-67, Lutetium-177 among them PSMA label with Lutetium-177 is most widely used agent
- PSMA also can label with alpha emitting radionuclide mainly Actinium-223

25.5.2 Treatment Requirement for Lutetium-177 PSMA Therapy [8]

- Patient must be a diagnosed case of metastatic castration resistant prostate carcinoma refractory conventional treatment.
- Current ^{68}Ga/^{18}F PSMA PET-CT scan show adequate PSMA expressing lesions
- Patient should not have liver metastases which do not show adequate uptake on ^{68}Ga/^{18}F PSMA PET-CT scan

25.5.3 Contraindications of ^{177}Lutetium PSMA Therapy [8]

- Life expectancy less 6 months or ECOG [9] performance status more than 2
- Unmanageable urinary tract obstruction
- Unmanageable urinary tract infection
- Dearrange renal profile

1. GFR less than 30 mL/min
2. Serum creatinine more than two times of normal lab value
3. Dearrange liver enzymes more than five times of normal lab value
4. Bone marrow suppression

- Total leukocyte count less 2500/mm^3
- Platelet count less than 75,000/mm^3

Impending spinal cord compression and pathological fracture.

25.5.4 Pre Therapy Patient Preparation [8]

- Recent blood investigations to rule out bone marrow suppression and derange liver or renal profile
- Recent ^{99m}Tc renal dynamic scan to estimate Glomerular filtration rate and any renal urine out flow obstruction
- Recent base Serum PSA levels
- On the day of therapy patient must be well hydrated and informed written consent is taken by the patient by informing about pros and cons of treatment and efficacy of treatment
  - Does of ^{177}Lutetium PSMA—100–250 mci of ^{177}Lutetium PSMA is given depending upon hematological profile, renal and liver function of patient
  - Therapy may give unwanted exposure other organs like bone marrow, kidneys and salivary glands
  - So maximum radiation toleration limits to red bone marrow, kidneys and salivary glands are 2, 28–40, 35 Gy respectively

25.5.5 Procedure

- 1–2 liters of normal saline is given at the rate of 20 cc/min to the patient for maintaining the good hydration.
- Slow infusion 100–250 mci of ^{177}Lutetium PSMA is given in 50 mL of normal saline.
- Diuretics and laxatives may be considered to eliminate unbound radiophramaceutical.
- Some patient may feel nausea during the therapy so antiemetics like ondansetron maybe given.
- In case of cerebral, spinal other painful or obstructive metastases corticosteroid should be start one day before the therapy and several days after the therapy.
- Cold pack may be applied on parotid glands during initial phases of radiopharmaceutical to decrease parotid gland uptake so risk of xerostomia is reduced.

After the competition of therapy post therapy scan is acquired under the gamma camera to confirm proper localisation of radiopharmaceutical at disease site.

25.5.6 How Many Cycles of ^{177}Lutetium PSMA Is Given to the Patient [8]

- 2–6 cycles are given to the patient at the interval of 8–12 weeks depending upon hematological, renal and liver function status

25.5.7 Follow Up of Patient After Therapy [8]

- After each cycle Blood cell count should be checked after every 2–3 weeks for up to 12 weeks
- According to PCWG3 criteria serum PSA levels should be check at the interval of 8–9 weeks for first 6 months from the initiation of therapy after 6 months serum PSA levels are checked after every 12 weeks
- Kidney and liver profiles are assessed at every 6–8 weeks interval
- After several cycles of therapy other radiological investigations are considered to detect PSMA negative lesions

25.5.8 Radiation Safety

- During first 2 days after the therapy high levels of radio pharmaceutical excreted in urine and patients are advised to observe rigorous hygiene to avoid contaminating the person using same toilet. A double flush is recommended after urination
- Patient should wash their hand after micturition
- Patient with urinary incontinence are advised prior to therapy and catheter should be kept for 2 weeks
- Urine bag should be emptied frequently. Gloves and protective clothing should be worn by staff for caring catheterized patient

25.5.9 Advantage of Lutetium-177 PSMA Therapy

- Since PSMA is overexpressed in most of prostate carcinoma and its metastases with respect to normal body tissue so it specifically target cancerous cell and normal tissues are spared
- It also shows rapid clearance from blood along with rapid and prolonged binding to cancerous cell so, high target to background ratio is achieved
- 30–70% of patients shows >50% reduction in Serum PSA level
- Studies show it can prolong the life expectancy

25.5.10 Limitations of Lutetium-177 PSMA Therapy

- Nearly 20% of metastatic CRPC patients have poor or absent PSMA expression so radionuclide cannot be given in these patients
- Some normal body organ like salivary gland, liver, spleen and kidney also show expression of PSMA, so unnecessary radiation exposure occur in these organs, dry mouth is common side effect
- Haematological toxicity also is a common side effect but can prevented by appropriate dose of radiophramaceutical
- Renal toxicity though can happen as a side effect but is not so common

References

1. Huang JF, Shen J, Li X, Rengan R, Silvestris N, Wang M, Derosa L, Zheng X, Belli A, Zhang XL, Li YM, Wu A. Incidence of patients with bone metastases at diagnosis of solid tumors in adults: a large population-based study. Ann Transl Med. 2020;8(7):482.
2. Wollin DA, Makarov DV. Guideline of guidelines: imaging of localized prostate cancer. BJU Int. 2015;116(4):526–30.
3. Jadvar H, Calais J, Fanti S, Feng F, Greene KL, Gulley JL, Hofman M, Koontz BF, Lin DW, Morris MJ, Rowe SP, Royce TJ, Salami S, Savir-Baruch B, Srinivas S, Hope TA. Appropriate use criteria for prostate-specific membrane antigen PET imaging. J Nucl Med. 2022;63(1):59–68.
4. Farolfi A, Calderoni L, Mattana F, Mei R, Telo S, Fanti S, Castellucci P. J Nucl Med. 2021;62(5):596–604.
5. Bouchelouche K, Choyke PL. Advances in prostate-specific membrane antigen PET of prostate cancer. Curr Opin Oncol. 2018;30(3):189–96.
6. Handkiewicz-Junak D, Poeppel TD, Bodei L, et al. EANM guidelines for radionuclide therapy of bone metastases with beta-emitting radionuclides. Eur J Nucl Med Mol Imaging. 2018;45:846–59.

7. Manafi-Farid R, Masoumi F, Divband G, Saidi B, Ataeinia B, Hertel F, Schweighofer-Zwink G, Morgenroth A, Beheshti M. Targeted palliative radionuclide therapy for metastatic bone pain. J Clin Med. 2020;9(8):2622.
8. Kratochwil C, Fendler WP, Eiber M, et al. EANM procedure guidelines for radionuclide therapy with ^{177}Lu-labelled PSMA-ligands (^{177}Lu-PSMA-RLT). Eur J Nucl Med Mol Imaging. 2019;46:2536–44.
9. Oken MM, Creech RH, Tormey DC, Horton J, Davis TE, McFadden ET, Carbone PP. Toxicity and response criteria of the Eastern Cooperative Oncology Group. Am J Clin Oncol. 1982;5(6):649–55.

GPSR Compliance

The European Union's (EU) General Product Safety Regulation (GPSR) is a set of rules that requires consumer products to be safe and our obligations to ensure this.

If you have any concerns about our products, you can contact us on ProductSafety@springernature.com

In case Publisher is established outside the EU, the EU authorized representative is:

Springer Nature Customer Service Center GmbH
Europaplatz 3
69115 Heidelberg, Germany

Batch number: 10726135

Printed by Printforce, the Netherlands